D1625394

The Finest in the Land

THE STORY OF THE PETROLEUM CLUB OF HOUSTON

THE STORY OF THE PETROLEUM CLUB
OF HOUSTON

Gulf Publishing Company
Book Division
Houston, London, Paris, Tokyo

The Finest in the Land

THE STORY OF THE PETROLEUM CLUB OF HOUSTON

Jack Donahue

THE STORY OF THE PETROLEUM CLUB
OF HOUSTON

Copyright © 1984 by the Petroleum Club of Houston, Houston, Texas. All rights reserved. Printed in the United States of America. This book, or parts thereof, may not be reproduced in any form without permission of the publisher.

Library of Congress Cataloging in Publication Data
Donahue, Jack.
    The finest in the land.

    Includes index.
    1. Petroleum Club of Houston—History. 2. Petroleum industry and trade—Texas—History. I. Title.
HD9568.H8D66 1984   367'.97641411   84-4645
ISBN 0-87201-284-0

This is a special limited edition of 200 copies of which this is copy number _____ .

DEDICATION

This book is dedicated to Charles A. Rosenthal of Schlumberger Well Services and Judd H. Oualline of Getty Oil Company. Rosenthal, chairman of the Petroleum Club's Publications Committee, and Oualline, a member of the Board of Directors and liaison between the Committee and the Board, were in charge of the project from conception to completion.

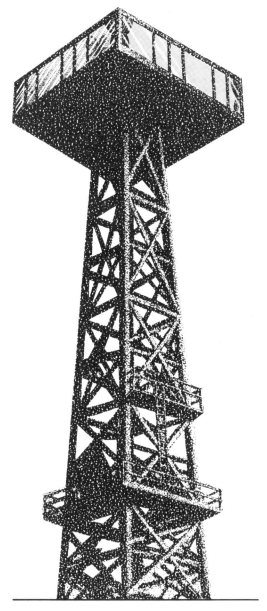

BY THE SAME AUTHOR

Someone To Hate
The Confessor
Erase My Name
Divorce American Style
Pray To The Hustlers' God
The Lady Loved Too Well
Wildcatter
Grady Barr (with Michel T. Halbouty)

CHAPTER ONE

ON A WINTER EVENING in 1945 two oilmen sat in their 17th floor office in the Esperson Building and watched the lights come on in downtown Houston. They were partners, content to rest a while in the warmth of their friendship before setting out for their suburban homes. It was a calming time, a brief reflective respite at the end of a tiring business day.

That morning they had been invited—urged—to become members of the Houston Club, a prestigious gathering place for the city's movers and shakers. Now one of them said, "You know, the Houston Club's fine, but I think this town ought to have a club for oilmen only."

His partner nodded.

"Everybody says Houston is the oil capital of the world," the first man said, "and it's ridiculous that we don't have an oilman's club. Why, there are thousands of us in this town! We've got . . . . "

His partner held up a halting hand. "I agree with you."

But the first man was already out of his chair and pacing the office floor. "It'll be the finest in the land," he said with certitude. "Everybody in the business will want to join, you see, and we can pick and choose."

The partner smiled, touched as always by the first man's contagious enthusiasm for any task at hand. But the idea of a club for oilmen was a good one, he thought, one worthy of pursuing.

1

"We can pick and choose where we'll locate the club, too," the first man continued, still striding back and forth across the office. "There's not a building in town that wouldn't be happy to have us." He stopped pacing abruptly. "Honorable men," he said. "We'll take in honorable men. Right?"

His partner got up and went to the door. "Come on. We'll get started on it the first thing in the morning."

The first man smiled. "It's going to be a lot of fun."

The Club *did* come into being—but only after six frustrating years during which the fun wore very thin indeed. And while a great majority of early members were honorable by any standard, the normal percentage of rascals managed their acceptance into the hallowed precincts. The rascals eventually were winnowed out, but some of them left colorful, lasting memories.

So did some of their more honorable brothers. Honorable men and rascals alike were of a generation of ready gamblers, happy to run a bluff or call one. Many of them had achieved maturity in rough and rowdy boomtowns of the 1920s and 1930s. A strong minority had drunk whatever was available. In heavy boots they had danced impartially with country virgins and shady ladies in noisy honky-tonks. They had seldom failed to lift a fist when challenged, and oftimes saw a challenge where none existed.

There were others, a majority, who didn't drink, didn't womanize, didn't fist-fight for the hell of it. But the rites of passage in a boomtown environment had toughened them mentally and physically nonetheless.

Quite a few of that generation—the wilder ones—had developed more than a nodding acquaintance with the fabled Texas Ranger Lone Wolf Gonzaullus in the brawling East Texas field, and more than one had spent a sleepless night on Lone Wolf's "trotline" or in the brand new Kilgore jail.

In a boomtown milieu such incidents were considered mere embarassments and not a black mark against one's character. The oilmen were young and full of themselves and wanted the world to know it. Lone Wolf was not averse to laying his pistol against their ears when they got too loud about it. He was a hard teacher in a hard school.

In 1956, some 20 years after the East Texas boom's heyday and five years after the Club's opening, he would visit the Club as a welcomed guest and marvel at its subdued opulence. A famed architect and a brilliant designer had pooled their skills to create what the early dreamers had prophecied would be "the finest club in the land. . . ."

But the Club would become more than a handsome watering hole. It would be said, perhaps in truth, that many of the decisions affecting the world-wide petroleum industry were made within its confines.

The early members were independent producers and those who provided them services. Later generations, however, were often representatives of giant corporations, and names of oilfields from Venezuela to Indonesia, from the Middle East to the North Sea, fell familiarly from their tongues. They varied from the old-timers not so much in substance as in style. There remained the rough-hewn honesty of a handshake being sufficient to close a deal, but an unending string of wars and political upheavals in areas once remote had taught a sophisticated caution. Raw personality edges, if they existed, had been beveled and polished by corporate politics.

But a down-on-his-luck wildcatter in good standing could still dine on gourmet food served by a perfect waiter while he tried to promote the drilling of a well in some God-forsaken spot where no well had been drilled before. In this haven he was the equal of any corporate executive eating at an adjoining table. And, under House Rule 3, neither of them was allowed to tip the waiter!

Houston was unique among American cities in that growth and prosperity had been the economic constants since the turn of the century. Even in the grim years of the Great Depression, when other population centers were cold and gray with closed factories, unemployed millions and appalling poverty, Houston was acclaimed in national periodicals as the country's "hot spot."

And it was, if only by comparison. No banks collapsed. Few businesses folded. Machine shops hummed. Refineries poured out streams of gasoline and other products. Great ships loaded in the ship channel. The staccato music of riveting hammers echoed in the downtown area in defiance of conventional wisdom. The whistles of railroad engines, lonely signals of expiring hope elsewhere in the land, were songs of promise in the Houston night.

Why was this?

Before the turn of the 20th century Houston was an agricultural community, a trading and distribution center, thriving chiefly because it sat at the head of tidewater on Buffalo Bayou, a stream open for navigation by small craft to Galveston and the open sea some 50 miles to the east. Goods from the outside world came into Galveston and were boated upstream to Houston and thence by railroad and wagon to inland Texas. Cotton, hides, and other products made the return trip to Galveston and on to distant ports.

Houston's ambitious entrepreneurs yearned to enlarge the bayou to accommodate major shipping vessels. They talked about such a project incessantly, but nothing was done.

Then, on September 8–9, 1900 a terrible storm struck Galveston, completely inundating the entire island and leaving an estimated 6,000 to 8,000 dead in the ruin and rubble. Not a single structure went undamaged. It remains the worst natural disaster in United States history.

Galveston had been struck by hurricanes before, as had other Texas port cities, and hardly had the wind died down over the stricken city than a cry went up for a "storm-free deepwater port"—logically, Houston.

And four months after the disaster—on January 10, 1901—the first great American oil gusher roared in at Spindletop Mound, a knob of land rising out of the prairie four miles south of Beaumont, Texas and only 85 miles from Houston. While Beaumont, a town of 9,000, became a madhouse as one spectacular well after another came in, Houston's canny entrepreneurs set about skimming off the cream.

One day Houston's 40,000 inhabitants were going placidly about their business, the next the town was rick-rollicking along on the crest of a boom that was to continue into the 1980s. Just north of the city the Humble field was discovered in 1904, and during the next 3 decades 13 significant oilfields were brought in within a 50-mile radius of city hall.

Buffalo Bayou was deepened, widened, straightened to form a great ship channel where, as the chamber of commerce boasted, 17 railroads met the sea. By 1930, there were eight refineries located along the channel to process the crude oil from the fields, and tankers delivered the finished products around the globe. Huge machine shops had sprung up to manufacture the tools and equipment the industry demanded. Great oil companies had been formed.

Houston, indeed, was the "Oil Capital of the World." A rebuilt Galveston, meanwhile, languished in Houston's shadow as a tourist center and, for a while, the seat of an illicit gambling empire.

The Great Depression settled on the land in 1930, but in that same year an old wildcatter, Columbus Marion (Dad) Joiner, sunk his drill into what would be for decades the largest, richest oil pool on the North American continent. Derricks studded the landscape in five East Texas counties as big companies and an army of wildcatters strove for a share of the 6 billion barrels of high-grade crude that lay some 3,600 feet below the surface.

Dad Joiner's discovery set off the wildest sustained boom in history. East Texas towns like Kilgore, Longview, Gladewater, Tyler, and Henderson would flourish, but the first oil to leave the field went to Sinclair Oil Company's refinery on the Houston Ship Channel. It was shipped over the Missouri-Pacific Railroad in 13 tank cars carrying 10,000 gallons each. The shipper was Haroldson Lafayette Hunt, who would emerge as the largest individual producer in the giant field and one the world's wealthiest men.

Houston had been a prime beneficiary of Spindletop and the other Gulf Coast fields. In the 1930s it reaped a golden harvest from East Texas. And frosting was slapped on the city's economic cake on June 5, 1932 when a Houston wildcatter, George W. Strake, completed the discovery well of the Conroe field, a significant oil strike 5 miles southeast of the city of Conroe and only 40 miles north of Houston. Like the East Texas field, the Conroe field was serviced out of Houston.

The last vestiges of the Depression disappeared with the beginning of World War II. It would be said that the Allies floated to victory on an ocean of oil. Most of that oil was produced in Texas and Venezuela. Because German submarines lay in wait to sink tankers in the Gulf of Mexico and along the Atlantic coast, much of the oil reached the great refining complexes of the New York-Philadelphia area through two giant pipelines—the Big Inch and Little Inch—originating at the East Texas field and the Beaumont-Spindletop area. They were constructed in record time under wartime conditions, and every day of their wartime service they delivered almost 500,000 barrels of crude and finished products to the Eastern Seaboard.

In June 1945, with victory in Europe assured, *World Petroleum Magazine* reported: "Allied invasion of enemy territory in Europe would not have been possible, nor could it have been sustained, without the aid of the 'Big Inch' and 'Little Inch' lines. . . ."

The U.S. government built the lines with the oil industry supplying the expertise and providing the manpower the great task demanded. With the war over, the government was ready to dispose of them. The splendid prize fell finally to Texas Eastern Transmission Company, a newly-formed company controlled by George and Herman Brown of Houston whose Brown & Root Inc. contracting firm was one of the world's largest.

But it was natural gas, not oil, that Texas Eastern pumped eastward in the big lines. Disdained by oilmen for decades, valued only because it pushed oil to the surface, sold when sold at all to companies retailing it to residences in oil provinces, natural gas became recognized during the war as a safe, inexpensive boiler fuel for plants and factories. Demand for it soared, and wildcatters who previously had sought only oil now deliberately drilled for gas. And they found it in abundance in Texas fields.

There had been speculation that with war's end the demand for oil would slacken. But those prognosticators had not reckoned with Americans and citizens of other industrialized societies who had not driven a new automobile for half a decade nor an old one without government regulation. The U.S. government had maintained a price lid of $1.20 per barrel on oil during the war. Now oilmen could see $2 per barrel around the corner and, fate willing, $3 per barrel.

It was against this background, and under these circumstances, that the two dreamers on Page 1, Wilbur Ginther and Howard Warren, decided to form a club for oilmen in a city oil had built.

It was fitting that the first discussion between Ginther and Warren about the Club was held in the Esperson Building. Niels Esperson, for whom the building was named, had been a wildcatter, a rugged Dane who had built a fortune in the early 1900s. His first strike had been at Humble, north of Houston, where he had drilled three dry holes on a five-acre lease before bringing in a gusher.

His wife Mellie had worked in the field with him. After his death in 1922, she carried on his various enterprises and increased the family fortune substantially. The Esperson Building, a 32-story skyscraper, had been the tallest building in Texas at its completion in 1927. Mellie later had built a 19-story structure adjoining it, and named it for herself. She had told the architect, "Don't build it as tall as my husband's building." In the public mind the structures were as one—the Esperson Building, fronting Travis Street between Rusk and Walker Avenues.

The building teemed with oilmen—lease brokers, drilling contractors, producers, geologists, petroleum engineers, consultants of several stripes. Many of them ate lunch in the ground floor drugstore, and they gathered in small groups on the sidewalk in front of the drugstore all during the day to talk about oil like smalltown teenage boys discussing the mysteries of sex.

There was similar oasis at the Cotton Hotel, an ancient edifice on the southwest corner of Rusk and Fannin. The lobby was a haven for lease brokers who had no offices but had working arrangements with the hotel switchboard operators. The alert young ladies paged the lease brokers to the house phones as if they were summoning them from important board meetings.

On the sidewalk outside gathered drillers, roughnecks, and roustabouts. "Cotton Corner" was their hiring hall and pickup spot. As customary, on the job they talked about women; on "Cotton Corner" they talked about the job.

About once a month the hotel operator, R. H. (Bob) Moffatt, would sashay into the lobby and, with angry shouts, drive the lease brokers out like a farmer shooing chickens from a prized garden. He would plunge among the roughnecks, cursing them for discouraging female patronage of the hotel coffee shop, scattering them in every direction. Within an hour every member of the cast was back in his place, confident that his lease had been renewed for another 30 days.

The Esperson Building contingent of the oil fraternity obviously was more affluent than the Cotton Hotel group. Other affluent producers, small and large independent companies, and major corporations were housed in other downtown buildings, including the Gulf Building on Main Street between Rusk and Capitol Avenues, the tallest in the city at 37 stories.

Wilbur Ginther and Howard Warren were affluent. Partners in the firm of Ginther, Warren, and Ginther, (the second Ginther being Noble, Wilbur's brother) they had opened their office in the Esperson Building in 1934. They had found productive wells in several Texas fields, but they would forever remain wildcatters, always looking for some giant in the earth. Noble, who was a member of the Houston Club, was not immediately impressed with the necessity for an oilman's club.

Wilbur Ginther was a wisp of a man, effervescent as Alka-Seltzer, friendly as a television game show host, but a bulldog who fought to a bloody draw with the Alberta, Canada featherweight champion in 1921. In that same year he ran second in the mile race of the Provincial Championship Meet at Calgary. He was a golfer, swimmer, and horseback rider, and any boomtown bully who took him lightly because of his size and demeanor was in for a major shock. Ginther's left hook was a dream-maker.

Ginther was educated at the University of Illinois and the University of Texas in business administration. But after helping his wildcatting father drill too many dry holes, he went to work selling the United States Chamber of Commerce magazine. It was an impossible job in depressed times, but he was second in sales in his fourth and last year on the job . . . and he had survived. "I just tried to put one in every home, like the family Bible," he told his friends when he joined forces with his brother and Howard Warren in the oil business.

Howard Warren was a Renaissance Man. Like Ginther, he was ebullient. A story-teller and charmer of landowners whose leases he coveted, he was also a music lover equally talented on a farmer's piano or a cathedral organ. But it was as a geologist, educated by the University of Colorado, and as a geophysicist, trained by Cities Service Oil Company, that he co-founded the firm of Ginther, Warren, and Ginther in 1934. Like Ginther, he had a passion for finding oil and gas. Like Ginther, he had a natural longing to gather with his own kind.

Shortly before Christmas 1945, the two men began discussing their plan with others in the building during lunch and coffee breaks in the drugstore. They were surprised when two other tenants, both drilling contractors, said they had been entertaining the notion of forming a club. They were Harris Underwood, Jr. and Edmund Pickering, Jr. Before long a little group of 11 evolved. To better discuss plans, the group began meeting for lunch in the Lamar Hotel coffee shop. And finally, in early January 1946, the men decided to hold a formal meeting in proper surroundings to demonstrate their solidarity and the firmness of their intentions.

Underwood and Pickering arranged for a private room at the Houston Club. The meeting was held on January 10, 1946, "for the purpose of organizing a club, the membership of which should be composed of persons engaged in the petroleum industry and allied businesses."

Wilbur Ginther was out of town and could not attend. Present were Warren; Underwood; Pickering; Vernon Frost, oilman and rancher; Ford Hubbard, landman (lease broker) and lawyer; Frank P. Donohue, manager of the oil loan department at National Bank of Commerce; R. E. (Gene) Chambers, an attorney for the J. S. Abercrombie Company; A. W. Waddill, top salesman for Houston Oilfield Material Company; Bert H. Nasman, district manager for Continental Supply Company; and B. A. Killson, an insurance man with oil interests.

Pickering served as temporary chairman and Hubbard as temporary secretary. Hubbard's minutes said:

> "In the general discussion it was the consensus that such a club was a necessity in Houston; that it would be well received by the industry; that there was little question as to its ultimate success.

"The possibility of securing one or more floors in the City National Bank when finally constructed was discussed, as was the fact that the organizers should be kept to a small number until definite plans could be presented to various members of the industry.

"The undersigned agreed to present information to the next meeting to be held Tuesday, January 15, looking toward the incorporation of such a club. It was thought that the name of the club should be either Petroleum Club of Houston or Houston Petroleum Club. It was also decided that each of the ten members present should submit a list of ten prospective charter members of the club to be presented at the second meeting of the organizers. . . ."

The second meeting was duly held at the same place. Ginther was on hand to argue that the name of the club should be Petroleum Club of Houston. On a motion by Killson and seconded by Ginther, it was unanimously voted that the club be so named.

A visitor was James A. Elkins, Jr. of City National Bank. The bank was constructing new quarters, and the possibility of securing one or more floors as club headquarters had been discussed at the first meeting. (Elkins offered them little or no hope of the new building as a club site, but some time later he was accepted as a club member in good standing.)

Ford Hubbard was directed to proceed with the filing of a club charter with the State Department in Austin, and after considerable discussion it was agreed that the three necessary incorporators of the club would be Underwood, Pickering, and Warren.

Underwood, Chambers, and Killson were selected "to look for suitable quarters," a seemingly simple task.

As for finances,

" . . . a motion was made by Frank P. Donohue and seconded by H. C. Warren that the organizers. . . . *pay into the treasury $10 each to defray the organization expenses in sight. This motion was duly carried. All present paid the amount stated. This money was paid to Ford Hubbard who was to act as temporary treasurer.*

"A list of prospective charter members of the club, consisting of leading producers of oil and gas, was compiled. This list was to be typed and submitted to each of the organizers and possible additions made to it. It was decided that each of the organizers would then contact a portion of the individuals listed to ascertain whether or not they would be interested in supporting such an organization as was being formed. . . ."

So, at the end of two meetings the little band had $110 in the kitty, a list of prospective members, a belief that a club was a "necessity in Houston," that it would be "well received by the industry," and that there was "little question as to its ultimate success."

Such optimism normally was reserved for the drilling of an offset well in the heart of a proven field.

Ford Hubbard drew up the incorporation papers, borrowing liberally from the charter of the Houston Club. Pickering, Underwood, and Warren were listed as the incorporators and directors. "The corporation shall have no capital stock, and has no assets at this time," the papers said.

The "purpose clause" read:

> "This corporation is formed for literary purposes to promote social intercourse among its members and to provide for them the convenience of a club house."

Hubbard mailed the charter with an accompanying letter to the secretary of state of Texas on January 29, 1946. He concluded the brief note: "I am enclosing my certified check for $60.00 to cover the franchise and filing fees. *Should the cost not equal this amount, will you kindly refund the difference.*"

Hubbard received a letter dated February 11 from J. L. McGarity, Head of Charter Division, Department of State. It said: "We have received a proposed charter for the above mentioned club and are returning the same to you. We suggest that you re-draft the purpose clause of said charter under Section 9 of Article 1302 of the Texas Revised Civil Statutes, 1925. . . ."

Since most oilmen were and are inclined to equate government regulations with socialism, or worse, McGarity's letter provoked a brief but bitter discussion before Howard Warren halted it with a few strokes of his pen. When he was through, the purpose clause read:

"This corporation is formed to support and maintain a club for playing handball and other innocent sports."

It should be noted that "playing handball" won out over "bicycle riding" by a narrow margin.

Why Warren's editing of the charter was considered an improvement in Austin is not clear, but Hubbard received a letter dated February 23 from the secretary of state himself, Claude Isbell. It said: "Enclosed is a certified photostatic copy of the charter of the above named corporation. This will acknowledge receipt of your remittance of $60.00, $5.07 of which will be refunded you within the next few weeks. The statutory filing fee for the charter and the franchise tax to May 1, 1946, amount to $54.93. When we may assist you again, call upon us."

It was not until early April that Hubbard received a state refund warrant from Mr. McGarity. Though tardy, the warrant was for the correct amount.

Meanwhile, the band of brothers had held an election of officers, naming Underwood president, Pickering and Waddill vice-presidents, Donohue treasurer, and Warren secretary.

So they were in business, legitimatized by the state of Texas and the Parliamentarian's Handbook. All they needed was a home and several hundred members to pay the rent on it.

CHAPTER TWO

THE MOST DESIRABLE LOCATION for the Petroleum Club of Houston, the members unanimously agreed, was the roof of the Rice Hotel. Sitting at the city's heart, 17 stories high, the Rice was the city's grandest hotel, "Houston's Welcome To The World!" The landmark fronted Texas Avenue between Main and Travis Streets.

In years past the Rice roof had been a romantic garden offering an unsurpassed view of the growing city to couples dancing to the music of big name bands. Most of the Club members remembered the roof with nostalgic affection. Now, in 1946, the roof was nothing but a storage place. It should be a simple matter to lease it, the members reckoned, and rather cheaply at that. The owner should be delighted to have the "finest club in the land" on top of the Rice.

Owner of the Rice was the Commerce Company, a corporation controlled by Jesse H. Jones. Jones, a colorful politician once grandiosely declared, had "done as much for Houston as any Caesar ever did for Rome." Jones detested such hyperbole, but it was true that he had been involved significantly in dozens of civic ventures that had proved vital to the city's progress. He also had amassed a fortune from various enterprises, including the erection of 50 commercial buildings. Jones' National Bank of Commerce, looking as much like a cathedral as a counting house, was on the ground floor of *his* Gulf Building. It was a city showplace.

Jones normally conducted business in his nearby Banker's Mortgage Building. To reach him, however, one generally had to go through his right-hand man, Fred Heyne, Sr. Heyne was not a frivolous greeter, and his cool, dry wit was sometimes lost on anxious callers.

When Harris Underwood arrived at Jones' office, it was Heyne who listened to Underwood's explanation of the club's ambitious goals, and hopes of building spendid quarters on the Rice roof.

The Rice roof, Heyne said, would lease at 39 cents per square foot per month, as is. The Club would have to provide for improvements at its own expense.

Now 39 cents per square foot was the going price for ready-to-occupy *office* space in Houston. Club members had been expecting to pay about 10 to 15 cents per square foot for the Rice roof with Jones meeting at least half of the necessary construction and decoration costs.

Heyne's proposition was a hard blow. The little group felt that he had set the price so ridiculously high because he held not a shred of faith in their plans.

He never wavered from his position over the next four years. His son, Fred Heyne, Jr., was among the first to join the 11 founders of the new club, and was one of its strongest boosters. He tried to deal with the senior Heyne, and so did other members over the years, but the price and conditions remained exactly as those laid down to Underwood.

Meanwhile, the hunt for a home went on. They all participated, but Underwood, a petroleum engineering graduate of Texas A & M, walked the streets of Houston like a gawking tourist looking for souvenirs or signs of sin. He was thorough in his examination of locations and, in most instances, as demanding as a phony prince.

One of Underwood's reports to the membership, detailed on six neatly handwritten pages, concerned—of all places—the Cotton Hotel. He had examined all 10 floors of the venerable structure, and recommended a specific use for each floor. Annual rent for the *entire building,* plus annual taxes and insurance, he had estimated at $60,000. The cost to air condition the hotel, install new elevators, to paint, and "make the hotel modern" was estimated at $300,000. To decorate, $200,000. The total cash outlay, thus was estimated at $500,000.

Where would the money come from?

The membership discussed that question. They would need 600 resident members and 500 non-resident members, with the resident memberships costing $400 each, the non-resident memberships $200. That would produce $340,000. Then they would secure $15,000 long-term loans bearing one percent interest from the five largest oil companies in town, and $5,000 loans from 26 smaller firms, for a total of $205,000.

Memberships plus loans would equal $545,000; $45,000 more than the estimated cost to "make the hotel modern." The surplus was a hedge in case some companies didn't want to make loans, or membership goals were not met.

But what about operating costs?

Well, the Club could maintain itself on income from a bar, dining areas, and room rentals. The 600 resident members would pay dues of $10 per month, or $120 per year, for a total of $72,000 annually, and the 500 non-resident members would pay $5 per month, or $60 per year, for a total of $30,000 annually. Thus dues income would amount to $102,000 per year. Since rent, taxes, and insurance would cost only $60,000 per year, there would be $42,000 left in the kitty annually to pay the one percent interest on the anticipated loans and to retire the debt itself.

Only a group of unflagging optimists would have entertained such a lofty notion. The major oil companies had ignored them. They had not offered financial assistance, nor had they instructed their executives to join the club. Neither had many of the large, rich independent companies. Time after time club representatives had called at the big companies' offices, and time after time they had been politely but definitely rebuffed.

From the first the club members had hoped—perhaps had even believed—that the major companies would buy big blocks of memberships for their ranking officers. Such a harvest of members would have given the Club instant credibility and financial stability. But it did not come to pass. Other companies would buy company memberships in time to come—not in big blocks—in order to transfer the memberships if the original holders were transferred or retired. Employees of some major companies who would join the Club in coming years would do so on their own—and they would not be numerous.

On February 17, 1947, the little group sent out a letter to prospects offering a charter membership to those who applied by March 31, 1947. By the end of that year they had attracted 28 new members, but had lost three of the founding fathers—Gene Chambers, Frank Donohue, and Bert Nasman.

Yet they carried on, exuding good will and confidence like a gang of disk jockeys. They did not write off the Cotton Hotel project because its magnitude fazed them; they simply set it on a back burner and continued the quest, hoping for something better to turn up.

They tried to make deals with other hotels, with banks, downtown apartment buildings, office buildings with vacancies. Motions were made at meetings, seconded and carried, to lease such and such locations, but something always occurred to make the members change their minds at the last moment.

They *almost* leased space above the Old South Cafeteria near Fannin and Rusk. They *almost* leased the second floor of the Ben Milam Hotel on Texas Avenue near Union Station.

And then they made a tentative deal for space with Fred Heyne, Sr., not on the Rice roof but in the Lamar Hotel, also owned by Jesse Jones. They engaged an architect, Roy W. Leibsle, to implement plans to "alter, remodel, furnish and equip" the space for club use.

They prepared, and mailed out to prospects, an attractive three-color brochure containing Leibsle's drawing of the floor plan and artistic renderings of the various rooms. The art work was excellent and the accompanying text persuasive.

"After more than two years of planning, Houston, the Nation's oil capital, will soon have a Petroleum Club," the brochure said. "On October 1, workmen will spread their blueprints and begin the preparation of its quarters. Sometime in January, its doors will open . . . revealing one of the finest clubs of its kind in the world. Housing game rooms, private dining rooms, a ladies' lounge, check rooms, and a large main club room, the Petroleum Club of Houston is planned with an eye to beauty and appeal. . . ."

On a page entitled CHARTER MEMBERS OF THE PE-
TROLEUM CLUB were listed the following names:

*"Chas. S. Atchison (Producer), Val T. Billups (Producer), V. A. Brill
(Consultant), Hugh Q. Buck (Attorney), Frank Champion (Producer),
W. D. Dunnam (Producer), George H. Echols (Producer), William B.
Ferguson (Producer), J. M. Flaitz (Consultant), J. M. Frost, Jr. (Pro-
ducer), J. M. Frost III (Producer), Vernon W. Frost (Producer), N. C.
Ginther (Producer), W. L. Ginther (Producer), Walter L. Goldston (Pro-
ducer), J. G. Gratehouse (Producer), M. T. Grubb (Producer), Karl Has-
selman (Producer), Fred J. Heyne, Jr. (Lease Broker), Joseph Hornberger
(Consultant), Ford Hubbard (Lease Broker), B. A. Killson (Miscellane-
ous), W. Howard Lee (Lease Broker), Harry Leyendecker (Producer),
Douglas B. Marshall (Producer), Glenn McCarthy (Producer), W. T.
Moran (Producer), E. E. Pickering, Jr. (Contractor), Paul V. Raigorodsky
(Producer), E. M. Reed (Banker), Corbin J. Robertson (Lease Broker),
M. A. Rutis (Supply), Harris Underwood, Jr. (Contractor), A. W. Wad-
dill (Supply), Howard C. Warren (Producer), Wesley W. West (Producer),
Herbert E. Williams (Producer)."*

These 36 men attended meetings, debated and voted, and
sought new members like recruiting sergeants trying to fill quotas
under a threat of court martial. This cadre, as it were, was willing
to meet on short notice in bank offices, spare rooms in other clubs
and, on occasion, homes of members with tolerant wives.

Also in this group was George Kirksey, a public relations man
of local repute. Kirksey had been engaged on a professional basis
when it seemed likely that the membership would find a home
above the Old South Cafeteria in 1947. That move was aborted,
but Kirksey continued to attend meetings. He found precious little to publicize, but he was not reluctant to join in club discussions
on any matter, and his opinions and suggestions apparently were
valued. And, in December 1948, his name showed up on the club
roster as a member in good standing.

Workmen did not spread out their blueprints on October 1,
1948 to begin work in the Lamar Hotel as the brochure had promised, but the membership drive the brochure kicked off brought
the muster roll to 180 resident members and 36 non-resident
members by end of that year. This was a far cry from the hopes
and dreams that he had floated across the tables in the Esperson

Building Drugstore, but the years hadn't been all bad. They had produced an indomitable cadre, for one thing, and the 1948 membership drive had brought into the fold Wilbur Ginther's "bell-cow."

Ginther had been preaching the need for a "bell cow" since 1947 when it became obvious that no help or guidance was forthcoming from the large oil companies. Again and again he told his comrades, "If we're going to get off the ground, we've got to get a big name people will respond to."

It made sense, and particularly to Harris Underwood, Jr. He was not a charismatic man. In a blatantly materialistic city he had not struck it rich. He was a hard worker, dedicated to achieving the Club's aims, but pragmatic enough to agree that he was not Ginther's "bell cow." On several occasions he had proposed the election of new officers, hoping to pass the presidential torch, but each time his comrades had vetoed the proposal, insisting that they needed his leadership. So he had stayed on the job, helping to hold the struggling club together with the glue of his devotion.

In the search for a "bell cow" the first name that came to mind was that of James S. (Mr. Jim) Abercrombie. He already was an oilfield legend. He had solved the riddle of the baffling oil sands at Old Ocean, Texas to bring in a rich field where others before him had failed.

Many years earlier, after walking away from a well blowout near Liberty, Texas, he had knelt on a country road and in the dust had sketched the first practical blowout preventer. H. S. Cameron, who ran a Houston machine shop, began manufacturing the devices after Abercrombie's pattern in 1922, and Cameron Iron Works, controlled by Abercrombie, was to become one of the world's great oilfield service companies.

As if this weren't enough, Abercrombie had gained the snickering admiration of the oil patch because of what was labeled "an act of God" in the Conroe field in the 1930s.

Standard Oil Company of Kansas drilled an offset to an Abercrombie well. The Standard well blew out. Standard capped the wild well, but the gas that caused the blowout found release in horizons near the surface, causing the Abercrombie well to "crater." The well's Christmas Tree—the controlling device—fell over and broke in the widening depression.

Oil spouted from the broken casing like an old-time gusher. Sumps were quickly dug, dams erected. The oil was contained and pumped into a pipeline gathering system.

Under the Conroe field's "allowable" rules, wells were permitted to produce only 150 barrels of crude daily. Now 16,000 barrels a day were pouring out of the cratered well—and being sold at the then-current rate of 60 cents per barrel.

Abercrombie was getting $9,600 a day from the wild well's output!

He had not been negligent. And he felt that wading into the crater to attempt repairs was an undertaking too dangerous to ask of any man—and state officials tended to agree with his assessment. Abercrombie, as an oilfield wit put it, appeared to be willing to wait out the well until the hand of God throttled it.

Not so the other producers in the field. They, and Abercrombie, were members of the Conroe Operators Association, organized at the start of the boom to ensure that the field would be developed in an orderly manner. Now Abercrombie's wild well was producing as much as 100 wells were allowed to produce.

After months of negotiation, Abercrombie permitted the drilling of a directional well to the bottom of the wild well, and the flow of oil was shut off.

It was estimated that Abercrombie had sold as much as $2,000,000 worth of oil from the cratered well.

By 1947, the year he occupied club members' minds as a potential leader, Abercrombie was held in high regard wherever oil was produced. In Houston, he was considered an outstanding citizen, noted as much for his charitable works as for his business acumen.

Perhaps Abercrombie's reputation made them hesitant, for the members dallied at approaching him. Finally Ginther, Warren, and Pickering were delegated to sound him out. The trio called on Vernon Frost to accompany them; he was the Club's tower of strength at delicate moments. Frost had no idea that Abercrombie would be interested in their plans, but he did not want to disappoint his friends.

The four men went to Pin Oak Stables where annually Abercrombie ramrodded an internationally-ranked horse show for the benefit of charity. They met with Abercrombie, but failed to bring up the reason for their visit. Warren's notes said simply, "It was deemed an unwise thing to do at that time. . . ."

They never found a wise time. And Abercrombie never became a member of the Club in any capacity.

Perhaps it was just as well, for the man who eventually became their "bell cow" was a genial giant who feasted on challenges. He was extravagant in word and deed, compassionate but hot-tempered, an oilman's oilman, and a leader born.

His name was Robert E. (Bob) Smith.

Smith was brought into the Club in October 1948 by Underwood and Ginther. They were his sponsors. Accepted as members in that same time frame were David C. Bintliff, E. O. Buck, and Marlin Sandlin. Working with the charter members—and particularly with Attorney Hugh Q. Buck—these men would set the Petroleum Club of Houston on the path to becoming "the finest in the land."

At a meeting in the Chamber of Commerce Meeting Room on March 21, 1949, Smith was elected president of the Club. Val Billups and Virgil Brill, charter members, were elected vice-presidents; Marlin Sandlin was elected secretary; and E. O. Buck was elected treasurer.

On the board of directors was Harris Underwood, Jr., who had served as president for three arduous years. Joining him were charter members Vernon Frost, Fred Heyne, Jr. and Hugh Q. Buck. David Bintliff and Ralph Johnston, a respected producer, filled out the slate.

"Boys," exclaimed an exuberant Wilbur Ginther at the meeting's close, "we've got us a hell of a powerhouse lineup!"

Bob Smith was 54, and in the pride of his manhood. A *Houston Press* reporter described him as having "ash-blond hair like a lion's mane, shoulders like a fullback, and eyes as sharp as drilling bits and kind as the lights on a Christmas tree. . . ."

He was richer than blueberry cheesecake.

His father had been a Texas oilman, and Smith followed in his footsteps. He was working in Mexico when the United States entered World War I. Smith hurried to Texas to join up. He was huddling with a group of likely recruits when the recruiting officer noticed Smith's right hand. The two lower fingers were missing, the result of a youthful hunting accident.

"The Army won't take you, son," the recruiting officer said.

"I've got my trigger finger," said Smith. "And you can pick out the toughest guy in this bunch and I'll bust his butt for you."

"Sign your name," the recruiting officer said.

After the war, Smith worked in an oilfield warehouse at West Columbia, Texas for a year, then took off for Oklahoma. On a cold winter night, in a small-town honky-tonk, he heard a drilling contractor say, "I've been freezing to death ever since I got to this damn country. If I could sell my drilling rig, I'd head for sunshine tomorrow."

The next morning Smith located the drilling contractor. "You were drinking pretty good last night," he said. "Do you still mean it about selling your rig?"

"Damn right," said the drilling contractor. "Give me twenty thousand and it's yours."

Smith went to the bank in the small town. He told the banker he wanted to borrow $20,000 to buy a drilling rig.

"What have you got for collateral?" the banker asked.

Smith put his hand on his chest, "I've got this old leather jacket and a pair of gloves."

The banker smiled, "If that's all you've got, why did you come to me?"

"Because you're the only one in this town that's got any money," Smith said.

The banker laughed, "Relax, son, and tell me about yourself."

Smith told him. He got the $20,000 to buy the rig. He moved it to Corsicana, Texas, put it to work, and paid off the loan. Then he began leasing land and drilling his own wells. He bought more rigs. He struck it rich at Pierce Junction field, near Houston, and then in South Texas before moving on to the great East Texas field and Conroe where his findings made him a multi-millionaire.

In all these years he seldom turned his face from a frolic, and never from a fist-fight. To be sure, he was slower to anger as he grew older, but even in his fifties he was known to go into a fighting stance at an insult or in defense of friends less powerful than he.

(Whipping aside the barber's apron, he sprang from his chair in the Esperson Building barber shop and knocked out—with one blow—an obnoxious patron who had been creating a disturbance. And long remembered was a friendly slap-fest he had with Albert Plummer, an oilman with size and temper, in a local health club. By the time spectators tore them apart, Smith had a cracked and swollen nose and Plummer was trying to hide the agony of a fractured rib cage.)

Withal, Smith was so gregarious, so generous with the bounty he had wrung from nature, that it seemed as if his name was associated with civic and charitable projects across the spectrum. "A noble soul," the *Houston Press* called him in an editorial lauding Smith's good works.

He had married Vivian Leatherberry of Houston, a tall, striking beauty with sleek black hair. They made a handsome couple attending various social events, and they never missed a local prizefight where the fight crowd, recognizing the Smiths as its kind of folks, greeted their entrance with yells and whistles.

As if Smith's life was not full and rich and satisfying enough, his willingness to gamble and his faith in a dogged leasebroker led him, in the late 1940s, to a treasure trove of almost incalculable value. The great find would become public knowledge in 1949, the year Smith assumed the Club's presidency.

In 1946, Humble Oil & Refining Company of Houston, one of the domestic giants, abandoned leases on some 4,000 acres in Scurry County, Texas after drilling a dry hole in the Ellenberg lime at 8,100 feet. To abandon acreage considered barren was a normal practice, but in this case a faint whisper drifted back to Houston that Humble crews, without knowing it, had drilled through an oil-bearing spongy reef between 6,300 and 6,700 feet. It was not a rumor to be readily believed.

But one who heard the whisper, and acted on it, was Homer Head, one of the many lease brokers who made their "offices" in the Cotton Hotel. Head's career had been marked with few successes and many failures, but not because he lacked diligence and imagination. The hotel's bellhops had little use for what they called the "leasehounds" who cluttered up the lobby. But they had respect for Head. "He works while the rest of 'em snooze," was a typical bellhop appraisal. And indeed he did. He was in and out of the lobby all day long, pursuing oil producers he tried to interest in drilling on leases he considered good prospects.

But Head was a student, too. He had no formal education in geology or petroleum engineering, but he spent his spare time poring over scientific tomes in the hotel lobby and any place he stopped to rest. He read current drilling reports instead of the *Daily Racing Form.*

When Head heard the whisper about the Scurry County dry hole, he sought—and found—the whisper's source. What he heard made him accept the rumor as truth. He went to the land owners who had leased to Humble. They were dispirited by the dry hole, but yes, they said, they would be willing to lease again.

Back in Houston, Head approached Hugh Roy Cullen, a famed wildcatter of immense wealth. Cullen was not interested in acreage where a dry hole had been drilled. So Head went to Bob Smith.

Smith listened. He examined Head's maps and papers. But chiefly he listened. He knew Head. He knew the lease broker had never scored a major *coup.* But he also knew Head was a hard worker and, to Smith's knowledge, Head had never dishonored his name.

"Get me the acreage and I'll drill it," Smith said, and they shook hands on the deal.

Head assembled 4,020 acres. Smith leased the land for the incredibly low price of $5 per acre, plus the customary one-eighth royalty to the landowners. Under the handshake deal with Smith, Head was to receive a one-thirty-second overriding royalty.

The deal was consummated in June 1948. On November 21, Standard of Texas drilled into the spongy reef on acreage not far from Smith's leases. The well flowed 528 barrels of oil during the official 24-hours testing period through a ¼-inch choke. It "proved up" Smith's acreage, and set off the biggest oil boom since the 1930s. Smith, ahead of the pack, leased acreage in every direction, but still there was plenty left for the rest of the boomers. Derricks dotted the landscape like pimples on an adolescent.

When the smoke cleared a bit, the Kelly-Snyder field, as it was called, came to view as a tremendous reservoir containing 1¼ billion barrels of the finest crude in the country. And Bob Smith was sitting smack-dab in the middle of it.

A *Houston Press* headline story on January 2, 1950, written by Ben Kaplan, began:

> "The biggest oil boom on the continent may make R. E. (Bob) Smith the biggest independent oilman in the world, and it couldn't happen to a nicer guy. If the Scurry County oil boom pans out, this Houston guy named Smith stands to make a cool one-half billion dollars. That would put him on the top of the heap among independent oilmen. . . ."

(The boom *did* pan out. By the early 1980s the field had produced more than a billion barrels, and another quarter-billion was waiting to be taken.)

The boom would make Homer Head a millionaire. It would lift him out of the Cotton Hotel lobby and into the Petroleum Club.

Smith had joined the Club after he had obtained the Scurry County leases and before the discovery well was drilled by Standard of Texas. He became Club president just as the immensity of the field was being determined. He had been a "bell cow" of the first order before Scurry County, one capable of attracting others to follow his leadership. Now he had hung another clapper in his bell.

Marlin Sandlin, at 40, was executive vice-president of Woodley Petroleum Company, a large, aggressive independent oil company controlled by Jubal R. Parten. Sandlin, Parten said on occasion, could do more work in half a day than most men could do in a week. Parten knew something about work. During World War II he had served as director of transportation in the Petroleum Administration for War, and he had driven himself and an army of men around the clock in the construction of the "Big Inch" and "Little Inch" pipelines from Texas to the Eastern Seaboard. It had been a job of Homeric proportion.

Sandlin also was a stickler for detail, an attribute Parten admired and one that could be appreciated by the Petroleum Club of Houston. Dark-haired and frank of face, Sandlin also was attractive, personable, and fastidious in dress. He had begun his adult life as an attorney with a love for politics. He has served as assistant secretary of state in Austin, and later had been legal counsel to the General Land Office. He had come to Houston as a partner in a law firm.

At a dance one night he had confessed to Wilbur Ginther that he was bored with his work. Many of the men on the dance floor with their ladies were in the oil business. Ginther had chuckled and waved a hand at the dancers. "The guys in the oil business aren't bored, Marlin. Their days are as much fun as their nights."

It so happened that Parten had lost a key man at Woodley Petroleum, leaving a gap in the land department and other areas. He had mentioned this to an old friend, James V. Allred, former governor of Texas. Allred had been impressed by Sandlin's work in Austin and with his political acumen. He recommended him to Parten, and Parten offered Sandlin a job at Woodley Petroleum. "Wilbur already had me conditioned, so I took it," Sandlin told a friend.

It was an easy matter, then, for Ginther to lead Sandlin into the Club and Sandlin, shortly thereafter, brought Parten into the fold. Parten told a friend: "Marlin works twenty-four hours a day for Woodley Petroleum, twenty-four hours a day for the Petroleum Club, and spends another twenty-four hours a day in being a loving husband."

The Club needed every ounce of his strength he could expend in its behalf.

A lot of knowledgeable people in the oil industry insisted that E. O. Buck, the Club's new treasurer, had played a major role in saving the great East Texas oilfield from destruction in the boom days of the 1930s. With his engineering know-how and his daring, they said, Buck had clearly demonstrated to oilmen and politicians alike that over-drilling and over-producing were wrecking the field's delicate reservoir mechanism. Buck, they said, had pointed the way to sane oilfield practices that eventually became the law of the land.

Buck had arrived in the field as a young engineer on a pipeline survey crew. He went to work for the Texas Railroad Commission for $230 a month. The Commission was authorized to "regulate" the field, to see that the field produced only as much as it was "allowed" to produce under the Commission's decisions. The Commission was ineffective, at best. "Hot oil," oil produced in excess of the Commission's "allowable," flowed out of the field in a veritable Niagara. New wells came in on the hour around the clock.

Buck persuaded the Commission to shut down the field while he tested the reservoir pressure with new-fangled devices called "Amerada bombs." The results were electrifying. The field's oil was driven to the surface by water that moved from west to east. In the west, before the field was shut down, bottomhole pressure was recorded by Buck at 1,400 pounds per square inch. From that point eastward, the pressure dropped steadily to be recorded at about 700 pounds per square inch in wells on the extreme east flank. Two years earlier a more primitive test had shown more than 1,600 pounds per square inch of pressure on the east flank.

It was evident that this drop in pressure was the result of oil being withdrawn faster than it could be replaced in the aquifer by the water drive.

Seventy-two hours after the field was shut down, Buck tested his key wells again. While there was no pressure change in wells at the field's west side—they remained at 1,400 pounds per square inch—the 700-pound pressure in wells on the east flank had risen to 1,300 pounds!

It was thus made obvious to the field's producers that they were drilling and producing the field to death.

Later on, when politics forced the Commission to increase the field's allowable, and the field was producing about half the total U.S. requirement, Buck flatly told the Commission that the reservoir pressure was dropping at an alarming rate. He warned that if the pressure dropped below 750 pounds it would be almost impossible to draw the inert oil from the producing sand even with pumps. Whatever was left at that point, he said, probably would be lost forever.

The Commission responded to Buck's warning, and cut back the allowable.

Buck had left East Texas to become chief engineer for the Conroe Operator's Association. He had helped see that the rich field was produced under the same sane rules he had helped promulgate in East Texas. He had helped negotiate with "Mr. Jim" Abercrombie in the shutting down of Abercrombie's "cratered well."

From Conroe he had moved to other assignments until Jesse Jones had tagged him to head the National Bank of Commerce's oil and gas division and shortly thereafter had made him a vice-president.

Frank P. Donohue, one of the original 11 members of the Club, had left his job at the bank and Buck had taken it. Donohue had moved on to New Orleans. Buck not only took over at the bank, he took Donohue's job as treasurer of the Club. Donohue, a *bon vivant*, had worked hard for the Club's success in his fashion. Buck brought to the Club a calm reasonableness. It would be needed.

At the time he joined the Club David C. Bintliff already was building a reputation as a financier and entrepreneur. He was a tall, slender, handsome man, a recognized super-salesman who, from 1933 to 1945, operated the largest American National Insurance Company agency in the country. Based in Houston, he had spread out into mortgage loans and real estate and, inevitably, into oil and gas financing. Many a Houston oilman got his start with money supplied by Bintliff. It was natural, then, for Bintliff to get directly in the oil business, and by the time he joined the Club he had considerable producing holdings in several oil provinces.

Bintliff was an intriguing mixture of suavity and "good ol' boy" charm. He had a sharp, inquisitive mind, with a talent for finding out what he wanted to know, and putting it to use if he so desired. He knew as much about Houston as the Chamber of Commerce, and apparently felt at ease wherever he found himself.

Bintliff, in short, was a diplomat. He brought his skills to the Club, and the members made him welcome.

Attorney Hugh Q. Buck—no relation to E. O. Buck—was a charter member. He had been brought into the Club by A. W. Waddill, a founding father. Buck represented Houston Oil Field Material Company where Waddill was the star salesman. Buck was a brilliant lawyer, a partner, at age 40, in the firm of Fulbright, Crooker, Freemen, Bates & Jaworski, one of the city's largest and best known.

While still in his early twenties, Buck had served as counsel to the Reconstruction Finance Corporation in Washington, D.C. The RFC was headed by Jesse Jones (he of the Rice Hotel, etc.), and Jones had given Buck a pleasant send-off when the young attorney decided to return home to Texas and to private practice.

But in 1939, when he was 30, Buck was called back to public service, being named to head the Utilities and Oil & Gas Division of the Attorney General's Office. Later he served as special legal counsel on utility matters to the cities of Houston and El Paso.

In 1943, he became a partner in the law firm of Gresham, Mc-Corquodale, Martin & Buck. That firm merged with Fulbright, Crooker, Freeman, Bates & Jaworski in 1945, and Buck distinguished himself in oil and gas matters.

Buck was physically imposing. He stood 6 feet, 4 inches tall, and weighed 212 pounds. When he lazily stretched himself to full height to deliver an opinion, he reminded club members of pictures of Abraham Lincoln preparing to launch an attack against Stephen A. Douglas.

It was not fear of Buck's bulk but respect for his brains and character that prompted even the most raucous members to silence when he had something to say.

Bob Smith, a keen judge of men, began relying heavily on Buck immediately after Buck was elected to the board of directors. Smith appreciated Buck's cool logic—and his willingness to take on any assignment for the betterment of the Club.

CHAPTER THREE

HENRY MARTYN ROBERT, who codified the rules of parliamentary procedure under the familiar title *Robert's Rules of Order,* would have approved most heartily of Marlin Sandlin. Whereas Howard Warren, the Club's first secretary, simply jotted down the minutes in the scantiest detail, Sandlin's reports were rich in fact and opinions. In his prolix paragraphs could be heard the thunder of debate and argument. They bristled with an importance befitting the minutes of a White House Cabinet meeting. Warren occasionally forgot to mention where a meeting was held. Sandlin apparently forgot nothing. It seemed almost that to him the minutes were more important than the meeting they reflected. Perhaps this is true of all inspired secretaries.

At the first meeting after the election of the new officers, Sandlin's minutes began:

> "BE IT REMEMBERED, that on this the 4th day of April, 1949, a special meeting of the Board of Directors of the Petroleum Club of Houston was duly and regularly convened at 2109 Second National Bank Building, Houston, Texas, at two o'clock p.m.
>
> "The meeting was called to order and presided over by the President, R. E. (Bob) Smith, as Chairman. On roll call, the following Directors were found present, to-wit; David C. Bintliff, Virgil

Brill, E. O. Buck, Hugh Q. Buck, Vernon W. Frost, Fred J. Heyne, Jr., Ralph Johnston, Marlin Sandlin, R. E. (Bob) Smith, Harris Underwood, Jr.

"Those present constituting a majority of the Board, the Chairman announced that the meeting was ready to proceed with the transaction of business. . . ."

Despite this pretentious beginning, little business was transacted. Some members had expressed unhappiness with the Lamar Hotel site, and economic feasibility had reared its ugly head. As a consequence, Hugh Q. Buck was instructed to inform Fred Heyne, Sr. that the club was withdrawing from the tentative deal. Buck "called attention to the fact that the plans for the construction of a new building by the Houston Club were well under way, and that, in his opinion, we should consider the use of a floor of such new building for the Petroleum Club. . . ."

David Bintliff "discussed the possibility of the purchase of a lot on Milam Street between Rusk and Walker with the view of constructing a building for housing the Club. . . ."

The Cotton Hotel as a club site was brought up again and apparently dismissed. Bintliff, Buck, and Frost were empowered to negotiate with the officers and directors of the Houston Club, but no one appeared enthusiastic about it.

It was just like old times. (Leibsle, the architect, never received all the money he claimed was due him for his work on the Lamar Hotel project but, succumbing to the soft sell of Howard Warren, did become a member of the Club.)

Bob Smith, however, made clear his position on the Club and its location. "Whatever decision we make should be with the view of providing a *permanent* location," he said, "and the facilities should be the very best available." Like Wilbur Ginther, Smith wanted the "finest in the land."

All through the summer and into the fall Bintliff, Buck, and Frost hunted a home much as Frost and Ginther and Underwood had sought one in earlier days. In October they met with other directors with news that they had negotiated with the trustees of the Hermann Estate for a lease on an apartment building at 801

Lamar Avenue for housing the Club. The contract needed only the approval of the board and the membership for work to begin. Preliminary plans for remodeling and decorating the premises already had been prepared by Thomas Bryan & Associates, Inc., architects.

More, one William T. Kent had been employed, subject to board and membership approval, to be the general manager of the Club.

The board voted to approve the lease and Smith and Sandlin were authorized to execute it, subject to action by the membership. The board also approved a contract with Kent; his job was to begin when the lease was executed.

E. O. Buck, the treasurer, spoke up. Applications for club membership had been pouring in. That was fine, of course. But it was going to cost a great amount of money to have the kind of club they all visualized. Therefore, said Buck, he moved that initiation fees for future resident memberships be increased from $500 to $750, and fees for future non-resident memberships be increased from $100 to $250, subject to the approval of the voting members.

Vernon Frost seconded the motion and it was unanimously adopted.

They had picked up some momentum now, and they didn't want to lose it. They ordered the calling of a special meeting of the voting members to be held on Wednesday, November 2, 1949, to consider the lease and building program "and such other matters as may come before the meeting."

In their euphoria, they voted themselves the luxury of obtaining the South American Room of the Rice Hotel for the session.

Wednesday, November 2, was only eight days away.

George Kirksey, the public relations man, had posters printed that pictured the property at 801 Lamar and described what could be done to it to make it a haven for weary oilmen. These were to be placed on easels at strategic spots in the South American room for perusal by members attending the meeting. A goodly crowd was expected to be on hand, and it was felt that Smith would have no difficulty in gaining approval of the project.

Then, on the afternoon of the meeting, Alwyn P. King, Jr., a partner in the oil business with Fred Heyne, Jr., saw some of the posters when he came into the office from an extended trip. King also was a club member. He pointed to a poster. "This shouldn't happen, Fred," he told Heyne. "That's a terrible place to build a Petroleum Club. Why don't you talk to your daddy again about the Rice roof?"

"It won't do any good," Heyne said.

"Then talk to Jesse Jones."

"You mean go over my daddy's head?"

"No. Talk to your daddy first. Tell him that you want to talk to Jones."

Heyne called his father and went to see him. The senior Heyne listened to his son's plea. He called Jones and asked him to see Heyne, Jr. "Send him in," Jones said.

Heyne, Jr. told Jones of the club's plans for 801 Lamar. Jones was aghast. "Not that old place," he said.

Heyne told him that club members had hunted a home for almost four years before Bintliff, Frost, and Hugh Q. Buck had found 801 Lamar. Jones saw Heyne to the door. "I'll handle it," he said.

Back at his desk, Jones called Bintliff, with whom he had done business, and asked him if he could come to Jones' office. Bintliff said he could.

About one hundred voting members were present when President Smith called the meeting to order and proceeded to bring them up to date on what the officers and directors had been doing since they were elected back in March.

Bintliff, it was noted by some, was not among those present.

But just as Smith was about to discuss the 801 Lamar location, Bintliff appeared. He made his way to Smith and Vernon Frost. He had just come from Jones' office, he told Smith and Frost. Jones had told him that he had built 801 Lamar 40 years ago, "and it wasn't any good then, and it's not any good now." Jones had said that he wanted to do something for the oil industry in Houston. He had asked for 15 days within which to make an appraisal of the Rice roof as a place to house the Club. He had said that he would reduce Heyne's asking rental price substantially and split the cost of remodeling the premises on a basis to be negotiated.

Smith relayed Jones' message to the membership. He thanked the members for their confidence and patience, and assured them they would be notified immediately when the officers and directors had a definite proposal from Jones.

His words brought joy to the hearts of the old guard who had pined for the Rice roof since 1946. Only the cynics wondered if Jones had another motive for his act. They could not believe that he had not heard of the four-year hunt for a club site; he had returned to Houston from his stewardship in Washington D. C. in 1946 to devote himself to his business. Little occurred in Houston of which he was not aware.

He was aware, for example, that on March 17, 1949 Glenn McCarthy, a publicized wildcatter, had opened his Shamrock Hotel far out in the boondocks at Main Street and then Bellaire Boulevard. The event had drawn international publicity and had made downtown businessmen wonder if the business district might not start creeping in a southward direction. Oilmen flocked to the hotel's Cork Club and dining rooms. The Shamrock, not the Rice, suddenly had become "Houston's Welcome To The World."

It could be, the cynics reasoned, that Jones now believed that the oilmen could, with their own club on the Rice roof, restore the aging hotel to its preeminent position.

In any event, Smith, Bintliff, Frost, and Hugh Q. Buck made an appointment to see Jones in his office at the Banker's Mortgage Building.

At 76 Jones was still handsome. He was tall, getting portly now. With his snow-white hair and warm eyes he looked like a benign uncle as he greeted the quartet from the Petroleum Club of Houston. He got them seated, then went back to his chair behind his desk. He laced his hands across his stomach. "All right, boys," he said.

Smith spoke. What the Club wanted, he said, was a 10-year lease and a 10-year option.

Before he could continue, Jones lifted a hand. "Boys," he said, "Jesse Jones doesn't *give* options, he *takes* them." (Many years later Vernon Frost would say that adhering to Jones' simple business principle had made him a lot of money in his business.)

So, there would be no option.

But they went to work, and the terms they agreed on were explained and discussed at a directors' meeting on November 28, and later on December 5 at a special meeting of the voting members in the Assembly Room of the Chamber of Commerce.

David Bintliff, as chairman of the Building Committee, told the voting members that Jones had offered to remodel and lease the east and center wings of the roof, comprising approximately 14,750 square feet of space, including a kitchen, on the following terms:

1. Rental would be $2,000 a month.
2. The Club would pay $125,000 of the estimated $350,000 remodeling cost with the balance to be borne by the Commerce Company.
3. The partitions, fixtures (except kitchen), interior decorations, and furniture would be installed at the Club's expense.
4. The Commerce Company (the Rice Hotel, in fact) would furnish a complete kitchen and waiter service to serve food to the Club with the understanding that upon a vote of 75 percent of the voting members, the Club could take over and operate the kitchen.
5. The term of the lease would be 15 years, and the Club would have the right to sub-lease with approval of the lessor.
6. Special elevator service for the Club's peak load period would be furnished by the lessor.

The voting members liked the deal, and authorized the directors to make it.

They also voted to increase the resident membership fee to $750, plus tax, and the non-resident membership fee to $250, plus tax, with the increases to apply to future memberships only. This was in accordance with the board's vote of a week earlier.

And, at long last, they finally came to a decision about monthly dues. They voted to fix the dues of a resident member at $15, plus tax, and the dues of a non-resident member at $5, plus tax, with a provision that $2.50 of each of the sums would be set aside as collected for a sinking fund to be used as the members might direct.

Hugh Q. Buck pointed out that resident members had paid only $100 down on their $500 resident membership fees, and it was time now to be calling in the balances due.

After some discussion, Fred Heyne, Jr. moved that January 15, 1950 be set as a deadline for balance due payments. Any member who failed to meet the deadline would be subject to paying a $750 membership fee, just as future members would be required to pay. The motion was seconded by Vernon Frost and unanimously adopted.

What about payment of dues? Val Billups moved that the monthly dues of all members start on January 1, 1950, payable in advance in quarterly payments. Virgil Brill seconded the motion and it was unanimously adopted.

Now they were in business.

At a subsequent meeting the directors voted to pay off William T. Kent, who had been hired as general manager when it was thought the Club would be housed at 801 Lamar. In his stead they hired Charles H. Boucher, who had just resigned as manager of the River Oaks Country Club. Office space for Boucher and a secretary was rented in the Rice Hotel.

A lively debate ensued when it came time to employ a designer. Marlin Sandlin strongly urged that the commission go to Edward J. Perrault, Jr., pointing out the River Oaks Country Club and The Fashion as examples of his work. Bintliff recommended employment of William John MacMullin, pointing out the Pine Forest Country Club and the Cullen Memorial Nurses Home as examples of MacMullin's skill. Both men appeared before the board of directors to vie for the job.

On a show of hands, Virgil Brill, E. O. Buck, Hugh Q. Buck, Vernon Frost, Fred Heyne, Jr., and Marlin Sandlin voted for Perrault. Val Billups and David Bintliff voted for MacMullin.

Harris Underwood did not vote. He had taken an instant dislike to Perrault and his design presentation. Some months later he resigned as a director and member of the Club. The directors accepted his decision "reluctantly and with regret," and thanked him for the valuable services he had rendered the Club. Though for almost five years he had striven for the Club's success, he never visted the Club while it was housed at the Rice Hotel.

They needed an architect to work with Perrault. A majority voted to employ Roy Leibsle inasmuch as he had worked on the proposed conversion of the Lamar Hotel as a site and was a member of the Club.

Hugh Q. Buck went to talk to Leibsle about the job. He came back to the next board meeting with a suggestion that someone other than Leibsle be employed. Leibsle, it had turned out, though happy to be a member of the Club, was still irate because the Club hadn't paid him as much as he thought he deserved for his work on the Lamar Hotel!

Buck said he had talked with Kenneth Franzheim, an architect who did a lot of work for the Commerce Company. Franzheim had recommended four firms, but had leaned in the direction of John F. Staub of the firm of Staub and Rather.

After some discussion Smith was authorized to employ Staub.

But Val Billups, Club vice-president, had been having some second thoughts about the plans for the Club. Billups, vice-president in charge of exploration for Texas Gulf Producing Company, had joined the Club in its infancy. He had worked hard for its success. But now Billups was worried about the country's economy, and he felt that a depression was a strong possibility in the near future.

It might be a mistake, he said, to make the expensive improvements contemplated for the Rice roof. Members might not be able to pay the monthly dues and the cost of eating at the Club. The Club could save money, he said, by buying out-of-stock furniture through local furniture stores. "We don't need a designer to handle that," he said.

It was obvious that Billups, like Harris Underwood, Jr., was unhappy about the employment of Perrault and with Perrault's tentative design presentation.

Billups was as tough as a mountain goat. He always spoke his mind, and forcefully. Now he implied that Bob Smith, with his wealth, was leading the directors and the membership to a plateau too high for survival.

For one of the few times in his life Bob Smith ignored a chip on another man's shoulder. The voting members wanted an outstanding club, he said. They had been promised one, and they were going to get one. As for the depression, "I feel certain that the affairs of the Club will adjust themselves along with everything else, and in some manner we'll be able to carry on without any regrets." He had spoken gravely.

Vernon Frost, Hugh Q. Buck, and David Bintliff quickly seconded Smith's stand. E. O. Buck went further. There had been too much talking and not enough acting, Buck said. It was time to get down to business. They had hired people to design a Club, so let them get to it.

Billups accepted the remarks of his peers with apparent good grace, but some time later he, like Harris Underwood, Jr., tendered his resignation. It was accepted with regret.

Once Perrault and Staub examined the Rice roof it became obvious that the east and center wings simply didn't provide enough space. The west wing was needed.

At a special meeting of the voting members on December 11, 1950, Smith explained the situation. He said that on his recommendation and personal guarantee as to the cost of the additional space, the directors had negotiated a contract with the Commerce Company for the entire Rice roof.

Monthly rental now was $3,500, not $2,000.

According to the minutes, Smith

"gave an overall estimate for the cost of the Club at $850,000, of which the Club's share will be approximately $500,000. The President further stated that the Club's opening date would be approximately June 1, 1951. . . ."

(As it turned out, Smith underestimated the cost by a substantial amount, and guessed wrong on the opening date by almost seven months).

The minutes continued:

> "The Secretary reported to the meeting that the Club then had 359 resident members and 113 non-resident members, and that there were 42 pending applications for resident memberships and 21 pending applications for non-resident memberships. The Secretary requested all voting members to submit names of prospective non-resident members to the Assistant Secretary and Manager, Chas. H. Boucher, immediately along with a letter recommending such persons for membership in the Club, *inasmuch as the directors had under consideration a program for securing non-resident members by invitation. . . .*"

What the directors had under consideration—indeed had acted upon—was the employment of a professsional member-getter, one Bob Lyles of Austin, Texas. By now the directors were convinced that 750 resident and 750 non-resident members were necessary to the Club's success. Lyles was to be paid a fee of five percent, plus expenses, on the net amount of money he produced by fetching in non-resident members.

E. O. Buck painted the financial picture. The club already had advanced $125,000 to the Commerce Company on general construction, and $26,011.54 for electrical work. There was $90,794.70 still in the kitty, but the cost to the Club for interior construction would be $212,871, and the estimated cost of furnishings and fixtures was $150,000. So they would owe $362,871 and had but $90,794.70 in the bank.

But, according to the minutes,

> "The Treasurer . . . pointed out to the meeting that considering the Club's resources from remaining resident and non-resident membership fees within the quotas of 750 resident members and 750 non-resident members, the Club will have a cash balance on hand after the Club's construction is completed. . . ."

The Club was short 349 resident members and 616 non-resident members. When we get them, Buck was saying, we'll be able to pay what we owe and have something left over.

If any member present considered this a ridiculous statement, he didn't say so. Indeed, the optimism in the meeting room was as palpable as the camaraderie. Certainly Buck, who never allowed his pragmatism to be seduced by a romantic strain, believed the goal was attainable or he would not have spoken with such tranquil assurance.

And Jesse Jones seemed to have no worries about the Club's future. He had been made a member, and perhaps his membership had infused him with the same rosy dye that brightened a wildcatter's vision. Jones had said on many occasions that he was content to make his money above ground, not below it. Few knew that early in the century Jones had been a member of the board of directors of Humble Oil & Refining Company. Having been persuaded to invest $20,000 in the new company, he had sold his stock eagerly after holding it for a year, bragging to his intimates that he had doubled his money on the gamble. (He would leave an estate of more than $250 million at his death; had he done no more in the oil business than hold on to his Humble stock, his estate may have rivaled that of Arabian princes.)

CHAPTER FOUR

THE RICE ROOF LOOKED as if a band of careless gypsies had been camping on it for a decade. To Staub the architect and Perrault the designer, it was a "huge trash pile," a "disaster area." Most of the roof was open to the weather, and winds had scattered debris in every direction. The part under a cover, with the exception of a small section housing a furniture repair shop, was jammed with hotel junk. It was a gray and ugly place, and the day was dreary. Standing there, with the wind rustling garbage around their ankles like leaves in a forest, the two men shut their eyes to the desolation and spliced their dreams of beauty.

John Fanz Staub enjoyed a deserved reputation in residential architecture primarily because of his work in River Oaks, an Edenic enclave populated by the city's wealthiest and most powerful citizens. To live in a "Staub home" in River Oaks was the Houston equivalent of pitching one's lamé tent on God's front lawn. Starting in the 1920s and continuing into the 1950s Staub drew up plans for more than forty River Oaks residences. They were stately mansions, for the most part, but there was a poetry about them as if each structure was a verse in a lovely song. "His refined style," an admiring realtor wrote, "was responsible for the graceful ambience we enjoy in River Oaks today."

Staub was 58 and at the height of his creative power when he undertook the Petroleum Club assignment. He had been reluctant at first, insisting that he needed no interior designer, that he would do the complete job himself. After a meeting with Perrault, however, he changed his mind. Perrault, he would tell a club member later, "is a genius, and a star salesman."

It was easy to understand Staub's initial reluctance. Perrault was a self-made designer with no formal education in the art. He had come to Houston at eighteen on New Year's Eve of 1928 from Jennings, in South Louisiana. Jennings was notable because the state's first oil well had been drilled nearby. Perrault was not interested in oil. Poor and broke though he was, he could trace his name back to Claude Perrault, who designed the East Facade of the Louvre for Louis XIV, the Paris Observatory, and decorations for the palace at Versailles. Claude's brother Charles was a controversial writer of the period, but is remembered best for his fairy tales—*The Sleeping Beauty, Cinderella, Little Red Ridinghood, Blue Beard, Puss In Boots,* and others. So on his first job in Houston— usher at Loew's State Theater on Main Street—Edward Perrault took tickets at the door and hoped that some day he would match the accomplishments of his famed antecedents.

One of his first customers at the theater was Jesse Jones. "Your ticket, sir," Perrault said to Jones. "I don't need a ticket," said Jones to Perrault. "*Everybody* needs a ticket," said Perrault. He was saved by the manager who explained that Jones owned the building and came and went as he pleased. Jones patted Perrault's head in approval of his dedication to duty.

He went to work as a stock boy at The Fashion, a high fashion shop that occupied the four bottom floors of the Kirby Building on Main Street. Ben Wolfman, owner of The Fashion, soon was permitting Perrault to try his hand at every facet of store operations, including design. "He's got more than a flair for design," Wolfman told friends. "He's got a great talent."

Perrault, meanwhile, was spending his free time designing sets for the Houston Little Theatre, and creating sets for Margo Jones at the Houston Community Players. But he had to try on New York for size. New York was a city of soup lines, but he managed to get a job creating displays in a department store. And he made friends with a family that took him on a four-month's journey to Europe. He feasted his eyes and spirit.

When he came back to Houston and The Fashion, he had not been forgotten. Wealthy families that remembered his work in the local theatres began commissioning him to design debut balls and weddings for their daughters.

This type of work was interrupted by World War II. Perrault spent more than three years in military service where his talent was put to use in the design and construction of "day rooms" and officers' clubs for Air Force personnel. He commanded a crew of 56 skilled artisans whose work drew praise and promotions from the Air Force brass.

Out of uniform he found Ben Wolfman waiting for him to design the remaining seven floors of the Kirby Building for use by The Fashion. It was two-year job with Perrault spending much of his time in New York. His work was so outstanding that he was chosen to remodel the River Oaks Country Club. It was there that Marlin Sandlin, a member, grew to admire Perrault's talent; he recommended him without reservation to help create a Petroleum Club of Houston that would be the "finest in the land."

The Club's minutes do not reflect how Perrault lured Club directors and other influential members away from their notions of boomtown murals, derricks in convenient niches and crannies, and displays of other oilfield memorabilia. Perhaps the members' tastes were more eclectic than they would publicly allow. In any event, the designer took them to the mountain top and showed them the primal source of energy, the sun, and drew his concept for the Club around it.

Perrault's inspiration was Louis XIV's worship of the sun. The monarch was known as the "Sun King," and this brilliant motif would become the central theme of the designer's concept for the Club. He would engage the noted calligrapher and graphics artist Andrew Zoeke of New York to make the "signature cut" for a sunburst that would show up on everything from china to match covers and so dominate one club section that it would be called the "Sunburst Room."

He would call on Frederick P. Victoria, designer and antique dealer of renown, to find the treasures his concept demanded. And he would pull famed artist Seymour Fogel away from an exhibition of his works at the Museum of Fine Arts to create striking murals on walls of marble dust. From a crew of skilled artisans he would get workmanship beyond their normal powers.

He would blend all of this in such a fashion as to prompt Wilbur Ginther to comment, "that Perrault is running in oil."

Little else, however, was running so smoothly. The Club and Charles Boucher, manager, had come to a parting of the ways. A woman known to Boucher but unknown to Club directors had entered the Club's 17th floor office in the Rice Hotel and had scattered files and furniture with reckless abandon. Boucher hadn't liked this, and the directors had liked it even less.

So the Club was looking for a new manager. "He's got to be top-notch," Bob Smith said. "And we can't wait too long in finding him." Nevertheless, weeks passed without a candidate meeting the directors' demanding requirements. Then an erroneous tip turned up a choice prospect. The Club had hired the national auditing firm of Harris, Kerr, Forster & Company to audit the Club books, and one of the firm members told Marlin Sandlin that the best hotel manager in America might be available. He named Lynton Upshaw, manager of the historic Williamsburg Inn in the restored city of Williamsburg, Virginia.

The auditor was wrong. Upshaw liked where he was and what he was doing. He had no intention of moving. And that's what he told Sandlin when Sandlin called him from Washington, D.C. where he was conducting some company business.

Thorough as always, Sandlin had investigated Upshaw's abilities and character. Upshaw had been manager of the famed East Lake Country Club in Atlanta before World War II, and he had returned from service in the Navy to take over the Williamsburg Inn job. He was 40, a pleasant man of quiet charm who was as thorough and hard-working as Sandlin.

"Don't be hasty," Sandlin told Upshaw. "Let's meet and talk about it."

Said Upshaw, "We've never met, but you tell me how much money we're talking about and I'll tell you whether there's any point in either of us pursuing this."

Sandlin told him.

"I'm interested," Upshaw said.

A short time later Upshaw flew to Houston to be interviewed. Sandlin took him to the Rice roof where work was in progress. He showed Upshaw the view from each wing. "Mr. Upshaw," Sandlin said, "you can't imagine what's going to happen in this town. It's the coming place, and you very well could play a part in it."

Upshaw was impressed. Sandlin took him to lunch where he met Smith and other officers of the Club. Later he was interviewed by the recently-formed House Committee composed of Hugh Q. Buck, Sandlin, Bintliff, and two early members, M. A. Rutis and William J. Goldston.

Rutis was blunt. "Mr. Upshaw, we know how much you've said you want. How much are you making now?"

Said Upshaw, "I don't mind telling you. You could find out elsewhere if you want to know badly enough. But *you* tell *me*, what's it got to do with the job you want done here?"

Rutis nodded. "I withdraw the question. What can you do for the Club?"

Upshaw detailed his experience. "I think I can run the Club in a manner to make you proud to be a member," he said. "If you want a Hollywood production, that's not my cup of tea. If you want a first-class operation, I think I can provide it."

"I think you can, too," Rutis said.

So said they all, and the Club had a new manager.

Meanwhile, Hugh Q. Buck and Sandlin had drawn up the Club's bylaws and had written amendments to the Club charter. Both bylaws and amendments had been approved by the membership. Two new directors had been named to replace Val Billups and Harris Underwood, Jr. They were Leonard F. McCollum, president of Continental Oil Company, and Claude A. Williams, president of Transcontinental Gas Pipeline Corporation. McCollum also took over Billups' job as vice-president. McCollum had been brought into the Club by Bintliff, and Frost had brought in

Williams. The newcomers immediately took active roles in the Club.

And the directors smartly enough voted honorary memberships to Governor Allan Shivers, Senator Lyndon Johnson, Senator Tom Connally, and Congressman Albert Thomas. They also accepted as a member Hamilton H. Kellogg, spiritual leader of the Christ Church Cathedral Parish. Under a special membership category, Kellogg's membership fee was only $100 and his annual dues were only $60, but he was not entitled to stock participation, voting, or transfer privileges of any kind.

"Stock participation" and "transfer privileges" were products of the new bylaws. For example, the $750 initiation fee for resident membership entitled the resident member "to one share of stock; to be evidenced by a non-assessable stock certificate as hereinafter provided; provided, that if the proposed resident member acquires from others a share of stock, no membership fee shall be charged, but the transfer fee hereinafter provided shall be paid and the consideration to the selling member shall be no greater than the membership fee then prevailing. . . ."

Non-resident members, who had no voting privileges, were issued non-assessable certificates of membership in lieu of stock certificates.

Pursuant to those same bylaws, the board of directors "drew by lots for terms as follows, to wit: three for one-year terms, four for two-year terms, and four for three-year terms." Fred Heyne, Jr., Bob Smith, and Claude Williams drew one-year terms; David C. Bintliff, Virgil Brill, L. F. McCollum and Marlin Sandlin drew two-year terms; E. O. Buck, Hugh Q. Buck, Vernon W. Frost, and Ralph A. Johnston drew three-year terms.

So in the summer of 1951 the Club had a set of bylaws, a new manager, and the charter had been amended. (The purpose clause of the charter had been rewritten as follows: "This corporation is formed for the purpose of operating a handball, golf, and social club." The rewrite job seemed to satisfy both the Club and the state of Texas.)

Despite this progress and an influx of new members with their infusion of cash money, the Club's bank balance was in such bad shape that it couldn't pay John Staub when he tendered a statement for $27,454. Bob Smith offered to write Staub a personal check. Speaking for the other directors, E. O. Buck thanked Smith for his generous offer, but said he believed he could manage the Club a $100,000 line of credit with Jesse Jones' National Bank of Commerce. And he did—Staub was paid.

So, borrowing and spending like the liberal Democrats they professed to abhor, the directors pushed on toward completion of the Club for oilmen in the oil capital of the world. But the bylaws now said the percentages of the active membership "shall be drawn from the following eligible business classifications: Oil, Gas, and Sulphur Producers, Refiners and Transporters, 50 percent; Oil and Gas Drilling Contractors, 6 percent; Oil and Gas Equipment Supply Men, 10.6 percent; Oil and Gas Professional Consultants, 7 percent; Oil and Gas Service Companies, 6 percent; Oil and Gas Lease Brokers, 3 percent; Oil and Gas Lawyers, 3 percent; Royalty Owners, 3 percent; Bankers for Oil and Gas Industry, 2.4 percent; Miscellaneous, 9 percent."

These allocations of percentages would plague the Club and prompt sharp debates right up until opening day. Some would say the allocations were a far cry from what the founders dreamed around the Esperson Drugstore tables. Others would say, to hell with the dreams, we're facing reality, and reality says the industry is served by myriad elements. As for the bulging "Miscellaneous, 9 percent," they argued, it could be put to good use by prudent Club officers and, after all, the category was not graven in stone.

That the Petroleum Club of Houston was being erected on the Rice roof had not gone unnoticed by the Houston news media. Indeed, Marlin Sandlin had granted an interview to Louis Blackburn of the *Houston Press* while lease negotiations with Jones for the third wing were still in progress. Sandlin had shared enough of Perrault's vision even at that point to make the interview credible. Uncharacteristically, Sandlin had used words and phrases like "swank" and "the last word" in discussing the club-to-be, and they conjured up mental pictures that would not be erased until the Club finally opened.

Bill Roberts, a columnist for the *Press*, seemed to know whenever a nail was driven in the structure, and he kept his readers duly informed. The media recorded it when a 7-foot statue and some 12-foot sofas, too big to be loaded in elevators, were lifted by crane through a hole in an unfinished wall, and huge cherry-laurels for a club balcony were hoisted up the face of the hotel.

On April 20, 1951—eight months before the Club opening— Ann Holmes, the *Houston Chronicle* Fine Arts Editor, wrote a thorough and fascinating description of what the Club would look like after she studied Perrault's sketches at a breakfast given by the Club directors. Other reporters would continue to write more stories until it seemed there would be nothing to write about when Club doors would open officially. They called the Club a "million dollar" project, and it was not officially denied.

At a directors meeting on November 21, 1951, it was decided that the Club could and would open on December 15, that there would be a preview showing for the news media on the 12th and a grand preview for members and guests on the 14th.

With that out of the way, Hugh Q. Buck, according to the meeting minutes,

> "pointed out that the Club's cash position as of the date of the opening of the Club would not be adequate to discharge all of the Club's obligations incurred in the construction of the Club's quarters for the following reasons, to wit: 1) There would remain unsold approximately 300 non-resident memberships and approximately 30 resident memberships in the Club; 2) increased construction costs had been incurred through changes and additions; 3) costs of furnishings had increased over original estimates due both to more elaborate furnishings being selected and inflationary markets."

That's not all, added Upshaw the manager. Two years before the membership had voted that $2.50 per month out of dues paid was to go into a reserve fund. This hadn't been done. The Club had spent all the money it could lay hands on to pay for the construction program.

Whereupon the directors unanimously adopted a windy, five-paragraph resolution saying, in effect, that they would explain the situation to the membership at the next general meeting.

There was nothing to worry about, Bob Smith implied. "I don't know of any club that offers so much for so little. When the members see what's been accomplished in construction and facilities, they'll be happy to cooperate with us in formulating a plan to retire our indebtedness . . . particularly from the standpoint of increasing the quota of resident members."

Increasing the resident membership quota above the 750 would not be a small matter, and it was pursued further in a board meeting on December 12, just hours before the Club was to be opened for viewing by the news media. Months earlier the directors wisely had authorized Smith to appoint a 15-man Advisory Planning Committee composed of well-regarded members. The directors had consulted with the committee from time to time during the Club's construction. Now Smith called the committee to this board meeting for advice about increasing the resident membership quota, and another matter he deemed most important.

Six members of the committee were old timers in that they had joined the Club before December 1948. They were William H. Hendrickson, vice-president of Texas Gulf Producing Company; Floyd Karsten, oil operator; Ralph Neuhaus, vice-president of Hughes Tool Company; Jubal R. Parten, president of Woodley Petroleum; Corbin J. Robertson, vice-president of Quintana Petroleum, Wesley W. West, who was active in oil, cattle, and lumber.

The remaining nine had joined the Club after Bob Smith's election to the presidency in March 1949. They were George R. Brown, vice-president of Brown & Root, Inc.; Donald L. Connelly, president of Warren Oil; John C. Flanagan, vice-president of a United Gas division; Archie D. Gray, Gulf Oil Corporation attorney; Pierre Schlumberger, president of Schlumberger Well Surveying Corporation; George W. Strake, discoverer of the Conroe Oilfield; R. D. Buck Walton, vice-president of American General Investment Corporation; Harry C. Webb, assistant to the vice-president of Texas Gulf Sulphur; John T. Jones, Jr., president of the *Houston Chronicle*.

After considerable discussion, a consensus was reached. The resident membership quota should not be increased unless justified through experience of operations for a reasonable period after the club's opening.

"What we should do," said R. D. Buck Walton, "is increase the cost of membership, not the quota. Both resident and non-resident memberships are too cheap when you consider what this club has to offer."

There was a general nodding of agreement, and after some more discussion it was decided unanimously to close the resident membership quota at 750 and to raise the fee for future resident members to somewhere between $1,500 and $3,000. The fee for future non-resident memberships also would be increased "to some reasonable figure."

Accordingly, the club secretary was instructed not to accept further applications for resident memberships "pending a definite policy determination."

Bob Smith had something else on his mind. The Club, he said, had outstanding debts totaling about $175,000. What he wanted to do, said Smith, was to go to the bank the next morning and personally pay off the indebtedness. That way, he said, the Club could open its doors "free and clear." It obviously was important to him that the Club start off with a clean slate; his love for the Club was evident in his voice and manner.

It was a typical Bob Smith gesture. It was not made to gain him praise—and everyone present knew it. Some of those present also could have paid out the $175,000 without missing it unduly. But they knew the club's obligations should be shared by all members in the end—and they told Smith so.

"Bob," said R. D. Buck Walton, "it's a fine thing you've offered, but a club like this ought not to have a 'Big Daddy,' and you don't want to have that name put on you. We love and respect you, and I'm saying this to you for your own good."

Several others echoed Walton's sentiment while expressing their thanks to Smith.

He gave them his brilliant smile as he nodded acceptance of their judgment and entertained a motion to adjourn.

More than one hundred fifty members of the news media attended the special "press showing" of the Club on December 12. The directors had spared no expense to entertain the visitors, providing an unlimited amount of alcoholic beverages and dinner of prime sirloin steak and baked Alaska. Under the Club's contract with the Commerce Company, the food was prepared by Rice Hotel cooks and served by hotel employees. And alas, the steak was as tough as a driller's boot and the baked Alaska runny. And the waiters—splendidly attired in spats and uniforms designed by Perrault—served the food and drink like hash-slingers in a greasy spoon restaurant.

Fortunately, the guests were so impressed by the Club's beauty and the startling view of Houston's skyline that the poor food and service went unreported. But for Manager Lynton Upshaw, the evening was a fiasco. He had never liked the idea of having to depend on the Rice, or anyone else, for food and service. Now, as he walked smiling among the guests, lending a hand where one was needed, his stomach was roiled by a vision of endless days at the mercy of the Rice crew. Complaining, he sensed, would achieve little.

The service and the food—hors d'oeuvres this time—were no better, in Upshaw's opinion, at the grand opening on the evening of the 14th. But, as at the press party on the 12th, the magnificence of the Club muted criticism.

The *Press*, the *Post*, and the *Chronicle* treated the opening as a major news event. The *Chronicle* published a "Petroleum Club Extra" edition with a banner headline which said: "The Petroleum Club Opens." The subhead said: "Luxurious Quarters Unveiled." The story began:

> "Houston's newest claim to fame was unveiled Friday night as some 2,000 guests swarmed into the Petroleum Club, topping the Rice Hotel, for the club's official house-warming. It was an event of which those present can be proud. They can boast of having taken part in one of Houston's major events. . . ."

The guests stepped off the five elevators serving the Club into a 100-foot long loggia and found themselves gazing at the city's skyline through a floor-to-ceiling picture window . . . a view whose focal point was the dramatic Gothic shaft of Jesse Jone's Gulf Building.

Jones and his wife, and Bob Smith and his Vivian, were at the head of a receiving line composed of the other Club officers and directors with their ladies. Fred Heyne, Sr. and his wife also were in the receiving line.

The *Chronicle* story said:

> "The opening . . . was a dream come true for Mr. Smith and other
> leaders in the project to form a Petroleum Club here and establish
> it as the city's, and one of the nation's, finest. The dream of the
> oil men met a most hearty reception when presented to Jesse H.
> Jones, owner of the Rice Hotel, and he entered into the scheme
> with the greatest enthusiasm. The completed project probably is a
> source of as much pride to him as his entry into the writing field
> where his first book has been rated a best seller ever since it ap-
> peared. . . ."

Jones also owned the *Chronicle*.

Club members, guests, and the news media agreed that the
Club, indeed, was the "finest in the land."

Said one *Chronicle* story:

> "Physically, the rooms and corridors of the club appear to soar in
> space, so planned that a panoramic view of Houston is on magnif-
> icent display, day and night. Recessed lighting and spotlighting,
> controlled in the individual rooms, gives in some rooms the dra-
> matic effects equal to those on any theatre stage. Another New
> Yorker (in addition to Frederick Victoria), William Dietrich,
> worked on the special lighting of the club's murals and specific art
> objects like Kwan Yin . . . a heroic gilt-bronze figure of the Chi-
> nese goddess of virtue and mercy, seven feet high on a four-foot
> pedestal. Dating back to 1593, Kwan Yin formerly stood in the
> forecourt of the Winter Palace in Peking. . . .

> "Past the loggia and the Chinese goddess is the dramatic Celestial
> Lounge, developed with particular emphasis on space and free
> movement. It has its own sky—a blue recessed ceiling from which
> hang countless mobiles in abstract forms.

> "Another abstraction of the south wall of the room is Seymour Fo-
> gel's mural "Cosmos," based on the forces and grand design of
> outer space. At the center of the room is a seventeenth-century
> celestial globe, symbol of the oil industry's coverage of the world
> and suggestive of the planetary movements round the earth.
> Striking also is the floor of the room, oak planking with thin lines
> and sections of white African mahogany, like notes on a score of

music. Over the window-walls hang 1,700 yards of Italian silk draperies. Furniture in the room is of natural Burmese teak. . . .

On and on the stories ran. The main dining room presented

"an elegant picture of eighteenth century oak paneling, modern brass palm trees, a porcelain stove ten feet high made for Emperor Franz Josef of Austria, and a dramatic wall of antique mirrors . . . radial patterns in floor and ceiling—in keeping with the club's sunburst insignia—tie in the oil machinery and equipment motifs of the grill room (Rig Room), entered through two massive riftsawn white oak doors. . . .

The grill room was

"illuminated by Finnish brass pilasters pierced with tiny holes. Behind the semicircular bar of walnut and mahogany is another Fogel mural, synthesizing the tools of the oil trade. . . .

There were detailed descriptions of the library, the Barracuda Room, Chirico Room, Doodlebug Room, Wildcatter Room (the gaming room, described in the *Press* as "beautifully-appointed, with attractive poker chips—$1 variety [cream and white], $5 chips [blue and white], $25 [yellow and white] and $100 [black and white]. There also were some red and white chips numbered simply 25—for what purpose not explained. . . .").

Not ignored was the "Snooze Room," a tiny retreat with two daybeds and shower facilities. "If members get a headache thinking about all this opulence," said the *Press*, "they can retire to the 'Snooze Room' to catch their 40 winks of sleep. . . ."

The newspaper society sections were packed with stories and pictures. Maurine Parkhurst, the *Chronicle's* society editor, wrote:

"The luxurious appointments of the club provided a striking background for the beautiful gowns. Mrs. Jesse H. Jones chose for the gala occasion a long black crepe dinner dress, the front of the bodice being trimmed with an insert of black lace. Mrs. Frederick Heyne, Sr. wore a gown fashioned of mauve French lace, and Mrs. B. Magruder Wingfield was in a short red taffeta dinner dress ornamented with a rhinestone sunburst clip.

"Mrs. R. E. Smith wore a dress of hand-painted black lace over black net, and Mrs. R. D. Buck Walton's gown of turquoise blue pure silk brocade was adorned with an aquamarine pin. Mrs. Virgil Brill was gowned in a navy blue and crepe taffeta beaded in front with blue iridescent sequins. Mrs. John T. Jones, Jr. wore a short strapless black satin evening dress and black satin jacket, and Mrs. Vernon Frost's formal evening dress was fashioned of emerald green satin and trimmed with satin flowers of the same color.

Mrs. Hugh Q. Buck was dressed in chartreuse chiffon. A mink cape completed her ensemble. . . . Mrs. E. O. Buck's short plaid taffeta gown was worn with a navy jacket, and Mrs. David Bintliff was in mauve Chantilly lace. Mrs. George Strake was gowned in emerald-green satin. An ornament of pearls and rhinestones trimmed the front of the belt, and she wore jewelry to match. . . ."

There were guests from as far away as Venezuela and from as near as New Orleans and Beaumont. It was, as one newspaper heading declared, a "brilliant, gay reception."

A *Chronicle* editorial said unequivocally of the Club,

"It IS the world's best . . . a series of marvels from the elevator doors to the corners of the most remote rooms. There is literally nothing like it in scale, lavishiness, taste, and styling anywhere in the world. . . ."

Such praise from the local new media was not pure chauvinism. If the *Chronicle* had an indirect stake in the Club because of Jones, the *Press* and the *Post* had none—and they were equally laudatory. So were the radio and television stations.

Supporting their appraisals was one made by officers of the American Institute of Interior Designers, who were given a tour of the Club by local members of the institute. The officers were so impressed—"ecstatic," according to a *Press* columnist—that they immediately offered a full membership in their organization to Perrault, the self-taught maverick. Perrault in later years would hold the highest offices in the organization and receive its brightest honors.

Staub the architect also was generally applauded. However, the Club directors had become disenchanted with him because he had

gone off to Europe during what they considered crucial construction stages of the Club. There even had been discussion about not
paying him in full. But in the euphoria of the opening all was forgiven, and Staub paraded as a hero through the admiring throng.

For the press preview, Smith and Sandlin has sent invitations to
newspapers and magazines the length and breadth of the country,
and a surprising number had sent representatives in return. On
the whole, their stories were a little more restrained than the accounts in the Houston newspapers, but there was almost general
agreement that the Club was, indeed, a magnificent achievement.

The most bitter dissenter was a noted columnist who did not
come to the party but gained his impressions of the Club from a
12-paragraph story in the *Wall Street Journal.* He was Sydney J.
Harris, whose column "Strictly Personal" originated in the *Chicago Daily News* and was widely distributed. After several blister-
ing opening paragraphs, Harris got down to business:

> "Furnishings from a dozen different countries make up the club's
> chop suey decor, with a total value of abut $750,000. It sounds
> like a Hollywood director's illusion of aristocracy.

> "There can be no doubt, if any was remaining, that Texas is now
> going through the period of mercantile hysteria which the East
> and Middle West experienced many decades ago. Substantially
> all the 'new money' in the country is to be found among the oil
> fields.

> "And, invariably, a sudden accession of wealth brings along with it
> a monumental bad taste, and an almost infantile delight in lavish-
> ness for its own sake. It is the children, or grandchildren, of these
> millionaires who will conscientiously give their money to librar-
> ies, colleges, and museums.

> "Chicago, in its heyday, was the most wildly extravagant of 'new
> money' cities; today the wealthy are raising funds for the Art In-
> stitute (as they should be) instead of building gaudy monuments
> to commemorate the acquisitive instinct. . . ."

Copies of Harris' column were passed around by Club members like insinuative cartoons ground out surreptitiously on a
mail-room Xerox. The members pretended that the barbs did not
sting.

When the Club officially opened its doors for lunch on December 15, 1951, it had 737 resident members on the roster and 389 non-resident members.

It had been exactly six years since the winter evening when Wilbur Ginther and Howard Warren had conceived the notion of a club for oilmen only that would be the finest in the land.

Of the 11 founders only 5 were still members on opening day— Wilbur Ginther, Howard Warren, Vernon Frost, Ford Hubbard, and A. W. Waddill.

The other six had left Houston or had dropped their memberships for a variety of reasons. They were R. E. (Gene) Chambers, Frank P. Donohue, B. A. Killson, Edmund Pickering, Jr., Harris Underwood, Jr., and Bert H. Nasman. Nasman, however, would rejoin the Club in 1952.

So the remaining founders had the club of their dreams. So did the cadre of early-joiners, and those who provided the impetus to get the doors open. It was a grand acomplishment. It was "another world," according to the *Chronicle*. "Both to the directors and to Perrault, the first conception of the club was a place where they could be transported into a world of their own. Thus it was designed . . . looking out into space, and capturing a wonderful panoramic view of Houston from every angle. . . ."

For all its wonders, however, the Club still failed to lure the major companies through its portals. "The independents moved their club from the Esperson Drugstore and the Cotton Hotel lobby to the top of the Rice," it was said around town, emphasizing the general acceptance that the retreat had been built by and for independents and those who did business with them.

The industry in those days was called the "oil fraternity," but it was a nervous brotherhood. Since the early 1930s the independent producers and giant corporations had co-existed under an arrangement that had permitted both to prosper and present a solid front to the common enemy—liberal politicians and federal government regulators. In the Club's first decade, the arrangement would threaten to sunder in acrimonious public debate.

CHAPTER FIVE

COLONEL EDWIN L. DRAKE, "Father of the Oil Industry," was no more an army officer than Peter Pan was an astronaut. But in 1859, near the village of Titusville, Pennsylvania, on a stream called Oil Creek, Drake deliberately drilled a well to find oil—and he found it. It was certainly the first such venture in the United States, and probably the world. Its true significance was that Drake had shown how oil could be obtained in quantity at a time when people were crying for a cheap illuminant and America's expanding industries were in need of a dependable lubricant.

Drake had been a farmer, clerk, express agent, and finally a railroad conductor; he had never donned a military uniform or heard a shot fired in anger. While convalescing from an illness in New Haven, Connecticut, he met a promoter named James M. Townsend. One of Townsend's enterprises held title to some land on Oil Creek. Entrepreneurs had been gathering oil from springs and seeps in the area and selling it to refiners. The refiners made lamp fuel from the crude. It was in great demand because harpooners had thinned the ranks of the sperm whales and "coal oil" was smoky and unpredictable.

A Townsend associate, George H. Bissell, had seen illustrations of men drilling for salt and brine. Why not drill for oil?

Townsend sent Drake to Titusville chiefly because Drake had a free railroad pass. And Drake found mail awaiting him which was addressed to "Colonel Edwin L. Drake." It had been sent ahead by Townsend, who obviously knew how to make a mountain out of a mole hill.

But Drake drilled the well. His crew dug a common well hole to 16 feet. Water flooded the hole. Drake got some cast iron pipe and drove it 32 feet to bedrock. Then he began to drill. When the bit reached 69½ feet, the pipe filled with oil. Drake rigged up a pump, and the well produced about nine barrels of crude per day. It beat hell out of soaking an Indian blanket in a seep and wringing out the crude into a wooden tub!

The well set off a boom in Pennsylvania, and oil fever swept across the land—and around the globe—like Asian flu. Oil was found almost everywhere it seemed, from California to the Caucasus, from Texas to Teheran, from Kansas to Kuwait. Most of the oil in the U.S., it appeared, was in Texas, Oklahoma, California, Louisiana, and Kansas; but it was Texas that caught the public imagination. Texas and oil were as inseparable in the public mind as bacon and eggs. The Texas wildcatter was almost as romantic a figure as the cowboy. Fields were found in Texas from the Permian Basin to the Gulf Coast, from the Panhandle to the Rio Grande. And in 1930, as recounted earlier, came the discovery of the East Texas field, largest on the continent with 6 billion barrels of recoverable crude in the reservoir.

Oil, then, was considered a peculiarly American industry by Americans, and Texas was its heart. Adding shine to this picture were the Americans who went abroad to find oil for U.S. and foreign corporations alike. They were familiar figures in jungles, deserts, and snowfields, swaggering a bit under their load of "good old American know-how."

Oil was discovered in quantity in Russia in 1873; in Indonesia, 1885; Mexico, 1901; Iran, 1908; Venezuela, 1914; Colombia, 1918; Iraq, 1927; Bahrain Island, 1932; Kuwait and Saudi Arabia, 1938. Other Middle East provinces and several African nations were waiting in the wings for the drillers.

Five American-based international corporations grew out of the oil patches—Standard Oil Company of New Jersey, which would become Exxon; The Texas Company, which would become Texaco; Gulf, Mobil, and Standard Oil Company of California. So did two based in Europe—Royal Dutch-Shell, a Dutch-British combine; and British Petroleum, whose majority interest holder was the British government.

Royal Dutch-Shell had its beginning in Sumatra, Indonesia, in 1885 from a well with less daily production than Drake's. But it spread around the world, entering the U.S. with its Roxana Petroleum Corporation in 1912, later changing that name to Shell Oil Company. British Petroleum was founded by an English gold miner, William Knox D'Arcy, who had been granted a concession covering a half-million square miles by the Sheik of Iran. He drilled into an oil-rich formation near the site of an ancient fire temple. British Petroleum also spread around the world. (As late as 1969, with a huge stake in Alaskan oil, it would gain control of Standard Oil Company of Ohio in order to have an outlet for its Alaskan production.)

Strongest of the American-based corporations was Standard of New Jersey. In addition to its foreign interests, Jersey controlled a number of American companies, including Humble Oil & Refining Company of Houston, a giant in its own right. (Later these American companies would be gathered under a single banner in Houston as Exxon USA to comprise the American branch of Exxon, the international corporation, which headquartered in New York City.)

The international corporations, their American branches, and other large domestic oil companies that produced and bought crude, refined it, and marketed the products, were lumped together as "the majors" by the independents.

In Latin America, the Middle East, and the Far East, oil and minerals were the property of the state regardless of who owned the land. In those areas the majors obtained vast concessions from the government or, more accurately, the *jefe,* sheik, emir, or sultan who ran the government. The crude they extracted from these countries was not expensive to produce.

In the U.S., majors and independents alike had to deal with thousands upon thousands of land owners, not governments. This, and other factors, made U.S. crude comparatively expensive. The majors, however, sold their foreign crude in Europe and Asia at the price crude commanded on the Texas Gulf Coast. Thus, 10-cent crude from Venezuela, for example, could be sold in Germany at the much higher Texas price. And as foreign production increased, the crude began flowing to the U.S. to compete with domestic crude.

From its beginning the industry endured periodic gluts and periodic forecasts of shortages. A forecast of shortage in the mid-1920s so alarmed the U.S. government that the Congress created the "Depletion Allowance," an income tax measure that allowed the industry to keep, tax free, the first 27½ percent of gross income. The untaxed money was to be spent in exploring for new oil in the U.S.

Within a year the fear of a shortage was being washed away by a glut. The majors blamed it on over-production by the independents. The independents blamed it on the importation of foreign oil by the majors. The newly-formed Independent Petroleum Association of America (IPAA) commenced lobbying in Washington for a tariff on imports.

During the shortage, crude had been selling for $1.88 per barrel. With the glut, the price dropped to $1.10. And the glut became a positive flood when Dad Joiner discovered the East Texas field in 1930, the first year of the Depression. With unrestricted drilling around the clock in East Texas, the price of a barrel of crude dropped to 50 cents. Two new fields were opened in Oklahoma, and crude prices dropped to 20 cents. In East Texas the price went down to 10 cents and, during one period, the high-grade crude sold for two-cents a barrel!

For years major company scientists had declared the East Texas region barren of oil, and they had not been particularly impressed by Dad Joiner's modest discovery well. Not so the individual operators and small independent companies. They swarmed across the five-county area, leasing any spot of land large enough to hold a derrick. When the majors belatedly made their move, they had to scramble for the leavings or, in some instances, buy leases from independents. And because the independents drilled their leases as quickly as possible, the majors felt compelled to do likewise. Derricks stood leg-to-leg, and there were no dry holes.

Everybody blamed everybody else for the resulting chaos, and everybody was to blame. The Texas Railroad Commission held the power and responsibility to administer a 1919 statute which prohibited waste and required conservation of oil and gas, but its orders were generally ignored or overthrown in the courts. The governor placed the area under martial law, but even the "boy scouts," as the National Guardsmen were called, could not effectively staunch the flow of oil.

And all the while the price of crude rose and fell like a monkey on a string, climbing to as much as $1 a barrel when a Commission order restricted production, falling back when the order was overruled.

Conservationists cried out against the physical waste of a precious national asset, for physical waste there was. Others decried what they called "economic waste," claiming that unrestrained production in Texas alone could almost satisfy the total U.S. demand for oil.

The Congress responded to the IPAA's plea for a tariff on imports in June 1932, imposing an excise tax of 21 cents per barrel on foreign crude. It would prove to be a hollow victory.

Five months later, on November 12, 1932, the Texas Legislature, after a bitter fight and a close vote, enacted what was described by many as the definitive conservation bill. It included in its definition of waste "the production of crude petroleum oil in excess of transportation or market facilities or reasonable market demand," and it did not prohibit the consideration of economic waste in the regulation of production. Similar steps were taken in other oil states.

Key phrase in the new law was "market demand." To restrict production to market demand the Commission and its counterparts decided how much crude each field should produce each month, what fraction of the production should be allocated to each well, and how many days per month the wells should produce to achieve the monthly quota. The majors were the "market." They told the Commission how much crude their refineries would require each month, and the well allowables and the number of production days were set accordingly.

Proration, as the process was called, resulted in a curtailment of production and a rise in crude prices in the U.S. Production of inexpensive foreign oil continued unabatedly. It was sold outside the U.S. at the Texas Gulf Coast price. And increasing amounts of it were brought into the U.S.; with proration it was still less expensive than Texas crude, despite the tariff.

The independent came to accept proration as a fact of life. His wells were limited to producing only a portion of their potential each month, but he could console himself with the knowledge that he was selling his crude for a reasonable price to a willing buyer. And the better the price he received per barrel for his limited production, the more the majors were able to charge their customers outside the U.S. for their less expensive crude from Latin America and the Middle East.

Critics snorted that the *modus vivendi* was a negation of the free enterprise system, no more than legal price-fixing accomplished under the bright banner of Conservation. The weary survivors of the fray appeared to regard that assessment as mere nit-picking. The arrangement would survive intermittent gluts, shortages, wars, revolutions, managerial blunders, and political stupidities, and within its framework both independents and majors would continue the hunt for oil.

The Texas Railroad Commission, in its wisdom, devised a "discovery allowable" for the proration program. Under it a discovery well could produce 20 barrels of crude per day for each thousand feet of depth, plus 20 additional barrels. A 6,000-foot well, then, could produce 140 barrels daily. The "discovery allowable" applied to the first five wells drilled in the pay zone. They were allowed to produce their 140 barrels daily every day of the month for 18 months or until a sixth well was drilled into the pay formation, whichever came first. Then the wells were placed on, say, a 15-day production schedule at 60 barrels per day.

Independents and majors also maintained a common defense against federal government encroachment and congressional foes who continually attacked the depletion allowance. (Ironically, this measure, enacted to stimulate domestic exploration just a year before the "Great Glut," would for decades endure assaults by liberal politicians—only to be abolished at a time when an acute oil shortage imperiled the country's economy and its security as well.)

A slackening of demand predicted for the post-war world did not occur. When a surplus began building in the late 1940s, it was erased by the Korean War, which began in 1950 and was supplied primarily by the majors from their foreign sources. In 1951, the year the Petroleum Club of Houston opened its doors, the Iranians overthrew the Shah and nationalized Iran's oil industry. The majors retaliated by boycotting Iranian crude, forcing the shutdown of the world's largest refinery at Abadan and taking 610,000 barrels of crude daily out of production.

In the meantime, it was getting increasingly difficult for a member to get a tasty meal in the Petroleum Club of Houston. The majors made up the Iranian deficit with crude from Kuwait and elsewhere, and eventually the American Central Intelligence Agency and British Intelligence would foment a revolt that would put the Shah back on his Peacock Throne. And Lynton Upshaw, the Petroleum Club manager, would wage war against the Club's poor food and service until every member and guest who dined in the Club would swear the meal was fit for a Shah!

He was worldly-wise, accustomed to catering to an eclectic clientele, accustomed also to working with a concerned and conservative ownership, but Lynton Upshaw was quick to realize that his new bosses—the directors—were conservative only in politics, and his new clientele—the members—were as hard to please as Nero Wolfe. The most boorish member knew a proper steak when he bit into it, by God. Many others who had worked and reveled in South Louisiana and in foreign lands knew that a spoonful of gumbo was a mouthful of glory and a Snapper Pontchartrain a sybaritic splendor. These men were aware of the virtues of sowbelly and beans or a bowl of red because many had relished such fare in leaner times, but in the rarefied air of their own retreat they expected their appetites to be pampered.

Upshaw had been startled several months before the Club's opening when he had gone to Bob Smith's office in the Second National Bank Building to get approval of a $5,000 expenditure. Smith had interrupted his free-wheeling business to kindly, but firmly, tell Upshaw to handle the matter himself. He had smiled his wonderful smile and added the hoary cliché, "You stay out of the oil business, Uppy, and I'll stay out of the club-managing business." A $5,000 expenditure in Upshaw's past would have required the approval of a committee after long, deliberate study.

Just weeks before the Club opened, and after it had been examined by members, Upshaw had been approached by W. Stewart Boyle, an early Club member. "Uppy," said Boyle, "I want to engage the entire Club for a party sometime pretty soon."

Upshaw had raised his brows.

"Oh, don't worry," Boyle had said quickly, "I'll furnish my own policemen."

Upshaw was quick to learn that squads of officers, arranged for through Captain Tom Sawyer, the "society cop," showed up as everything from traffic directors to protectors of furs and jewels at Houston social events . . . for a fee.

Nobody had to tell Upshaw about the quality of the food being served in the Club. He had been appalled by the food and service at the Club's media showing and the grand opening, and there had been no improvement in either after the formal opening. Many members had complained, one saying, "I can get food this good downstairs in the coffee shop."

The directors also had been fielding gripes. The problem came up for discussion at the very first board meeting after the opening. Hugh Q. Buck reported that he, Upshaw, and Smith had been making a careful study of the problems and were trying to work out a solution. "We think it would be best if the Rice continues to provide the food service, if it's at all possible," Buck said.

Several of the directors were dubious of such a course. Said Vernon Frost, "It's up to the directors to see that top quality food is served efficiently, and I mean on an individual basis as distinguished from a commercial basis, and we've got to do it regardless of what it takes."

Nothing was decided, however, and Upshaw began taking his case to Ben Orr, the Rice manager. Two or three mornings a week, Orr would arrive at his office to find Upshaw waiting for him. Orr's answer was always the same: the hotel was a profit-making institution and it was his job to see that a profit was made. Upshaw would call on the Rice chef, whom he considered a good one. But the chef would shake his head. "I take my orders from Mr. Orr," he would say.

But finally, in late March, Orr bent a little. "All right," he said wearily, "you pick somebody to work in the kitchen, and I'll pay the salary," and he quoted a price he would pay.

Upshaw hired Mrs. Elizabeth Reynolds, whom he had known in Williamsburg. She had been the food manager at the Travis House, a restored tavern, and before that had owned a restaurant in New York City. The directors also had the foresight to appoint a Food Committee with Herbert Seydler, a banker, as chairman. With Seydler's help, Mrs. Reynolds was able to make some minor improvements in the food and service, but after several frustrating months she threw up her hands and departed.

The service problem resolved itself when Upshaw, shortly before the Club opened for lunch one day, gathered the Main Dining Room staff and told them that after lunch there would be an hour of instruction for them.

"No, there won't be," said the dining room captain.

"Yes, there will be," said Upshaw.

The captain waved a hand. "Come-on," he said, and most of the waiters followed him out of the Club.

Upshaw called Ben Orr and demanded that the waiters be fired. Orr, without argument, discharged the waiters. Upshaw made do during lunch with waiters from other dining areas doing double duty. Then he placed an advertisement for waiters in the newspapers which emphasized that lack of experience was no handicap to employment. "We'll teach you," the advertisement said. Orr agreed with Upshaw's plan.

That evening Jesse Jones called Upshaw. There were three long-time Rice Hotel waiters in his office, Jones said. Ones that had been fired on Upshaw's insistence. "Why did you insist, Mr. Upshaw?" Jones asked.

Upshaw told him.

"You've answered my question," Jones said. "Thank you."

Club members waited patiently while Upshaw trained the new waiters the advertisement produced, but their complaints about the food were not diminished.

Finally, Hugh Q. Buck told his fellow directors that the Rice didn't *have* to do anything to improve food and service, that Rice management had lived up to the letter and intent of the contract between them, and that the Rice couldn't satisfy the Club's requirements within the policy of the main hotel food operations. "The quality we're demanding now exceeds the quality contemplated by the contracting parties in the early stages of planning our facilities and requirements," Buck said.

He said he and other members of the House Committee had met with Upshaw and Seydler and had studied every ramification of the problem. He asked Sandlin, a House Committee member, to report. Sandlin, in his neat, thorough manner, did so. It amounted to a recommendation that the Club take over the entire food preparaton and operation at its sole risk, cost, and expense. When asked, Upshaw said the risk was real that the Club could incur a loss in its annual operation, but he favored the move nonetheless.

That was enough for David Bintliff. He moved that Sandlin be authorized and directed to negotiate a deal with Fred Heyne, Sr. Ralph Johnston seconded the motion and it was unanimously adopted. Sandlin said he would act promptly, which brought smiles to the faces of the others. Had he ever done otherwise? He asked that Upshaw and Seydler help him.

Sandlin was back with a tentative deal in his pocket at a directors' meeting on December 10, 1952, to which members of the Advisory Planning Committee had been invited. Only 5 of the 15 members of the committee were on hand at the momentous meeting—George R. Brown, John C. Flanagan, William H. Hendrickson, Harry C. Webb, and John T. Jones, Jr.

To accomplish what it wanted to do, Sandlin said, the Club would have to take over 2,125 square feet of kitchen space on the roof at $2.487 per square foot or $5,284.87 per year. Also, the Club would take over and convert to its own use the entire north wing of the 17th floor of the hotel at an annual rental of $10,800. Alterations and purchases of Rice equipment would cost an estimated $113,389.75.

This meant, of course, that monthly rent would rise to $4,965.40 with all alteration costs to be paid by the Club alone.

Upshaw pointed out that the Club had paid its $175,000 note obligation to the National Bank of Commerce and was now free of debt. "To finance this project," he said, "a bank credit of $100,000 will be required. Such a loan can be retired through the sale of new memberships and the use of reserve funds. We would be optimistic to expect a profit from the food department the first year of operation. However, the potential for profit exists and, given time to establish a reputation, our volume of business should be enough to assure a reasonable profit."

It was a wildcat venture. So Leonard McCollum, a consummate gambler, moved that the Club borrow $125,000, not $100,000, and make the deal with Houston Endowment, which had taken over management of Jesse Jones' properties from the Commerce Company. Virgil Brill seconded the motion and it was unanimously approved.

It had been a year since the Club had opened to rave reviews with only the members and guests knowing that the food and service could be matched at a good Main Street diner. But with the remodeling accomplished, with his own chefs in the kitchen, and his own waiters on the floor, Upshaw would be confident that the Club's most demanding gourmet, Harry C. Webb, could leave his table with a happy palate while digestion's blissful tremors shook his portly frame.

Though their appetites appeared to dominate their thinking during the first year of operation, the members found time to increase the entrance fee for new resident members to $1,500 with annual dues of $240. And, with 75 potential members on a waiting list, the ceiling on resident membership was lifted from 750 to 900. And smartly enough, the three members of the Texas Railroad Commission—Olin Culberson, Ernest O. Thompson, and W. J. Murray—were made honorary members of the Club. And Bob Smith, in a private move, paid the entrance fees and dues for six clergymen of varying denominations.

After study by Fred Heyne, Jr. and Hugh Q. Buck, a hospitalization plan for employees was arranged. Employees also were voted a Christmas bonus. Upshaw was given a raise and a surprise Christmas bonus equal to a month's pay.

Upshaw by now felt at home among the Texans. He could recall with a smile an incident that occurred just after the Club opening. In his hearing, two new members who had not yet met him were looking over the Club. "What I don't understand," one of the men said, "is why they felt like they had to go up north and hire a yankee son of a bitch to run this club." A Georgian, Upshaw had spent his early working years in Atlanta and, as noted earlier, was lured to Houston from Williamsburg, Virginia.

The following year saw the end of Bob Smith's presidential tenure. He and Fred Heyne, Jr. and Claude Williams ended their terms as directors, but the membership reelected Smith for another three-year term. W. O. Bartle of Brewster & Bartle, drilling contractors, and William J. Goldston, vice-president of Goldston Oil Corporation, replaced Heyne and Williams on the board.

Marlin Sandlin became the new president, elected by the directors after a unanimous standing vote. David Bintliff and Vernon Frost were elected vice-presidents; W. O. Bartle, the new director, was elected secretary; Hugh Q. Buck was named treasurer, and E. O. Buck became assistant secretary-treasurer.

Several months in the past, Leonard McCollum and his wife had dined at the Club on an evening when no more than ten other couples had been present in the big, beautiful dining room. "We had a nice dinner," McCollum told Upshaw as he and his wife were leaving, "but where is everybody?"

Upshaw replied that it had been an unusually slow night.

"I don't blame folks for not coming here," McCollum said. "There's nothing going on."

He had brought the matter up for discussion at the next directors' meeting and it had been thoroughly discussed. Now, as president, Sandlin called attention to the need for an active entertainment committee. He suggested that Michel T. Halbouty, an earth scientist and independent producer, be named chairman of such a committee with the authority to select the other members. His suggestion was unanimously adopted. Halbouty and another independent producer, John Mecom, were named to the Advisory Planning Committee to fill existing vacancies.

Smith was not present at this meeting where he turned over the presidential reins to Sandlin, so the directors felt free to reminisce about his work for the Club. Finally, by unanimous action, the following resolution was adopted:

"WHEREAS, Mr. R. E. Smith, the retiring President of the Petroleum Club of Houston, has served the Club faithfully for four years as its President; and

"WHEREAS, the fact that the members of the Club are privileged to enjoy the most unique and outstanding club facilities of their kind known to exist anywhere in the world is due largely to Mr. Smith's generous and unselfish support of the Club's construction program; and

"WHEREAS, the aims and objectives of the Club have been well established on a conservative and sound basis, and the Club has enjoyed competent management and financial stability under his leadership; now, therefore, be it

"RESOLVED, that the Directors of this Club express to Mr. R. E. Smith their appreciation for his generous support of and his unselfish devotion to the Club and his outstanding leadership as the Club's chief executive during the past four years.

"RESOLVED FURTHER, that the Secretary furnish a copy of this resolution to Mr. Smith."

The resolution was warm with camaraderie and would be cherished by the recipient as long as he lived. But it is likely that the tribute that pleased him most came from a fellow oilman in March of 1949 when it was announced that Smith had been elected president of the Club. "Bob," said the oilman, "if you're going to be president of that damned club I've been hearing about, I'm going to join it in the morning!"

Years later Smith told a *Houston Press* reporter, not for publication, "I had been a lot of places and done a lot of things, and I had a pretty good opinion of myself, but when that man—a man I admired—said what he said, he made me feel that I was worthy of leadership. So I made up my mind to be a worthy president."

The early members of the Club had wooed him because they desperately needed a man of his stature to get them "out of the wind and cold and into a warm nest," as Howard Warren had described it.

Smith had done that, and more. He had been worthy, and more.

CHAPTER SIX

Bob Smith had been a powerful force, but not an intimidating one. Marlin Sandlin slid easily into the presidential chair and immediately began building on the strong foundation Smith had shaped. But Sandlin built in his own fashion. He was a delegator. He believed that the Club could best be served by strong committees headed by strong chairmen. He appointed tried-and-true old hands to head the committees, but he saw to it that the committee members were men capable of leadership. The officers and directors would continue to make the final judgments, but he wanted the committees to exercise initiative, and they did their work with the knowledge that Sandlin expected much from them.

He told his intimates that the circle of leadership should be widened, that active, dedicated committee members would provide a continuing supply of officers and directors. In this same vein, Sandlin recommended and gained approval of changes in the by-laws that prohibited an officer from serving a second consecutive term in the same office and a director from succeeding himself for a second consecutive term.

Hugh Q. Buck was chairman of the House Committee; Leonard McCollum, Finance Committee; Michel T. Halbouty, Entertainment Committee; H. M. Seydler, Food Committee; A. L. Selig, Library Committee; George T. Barrow, Publications

Committee. Selig, a geologist, and Barrow, an attorney, were both early Club members.

Bob Smith, it will be recalled, appointed an Advisory Planning Committee while the Club was under construction. The committee stayed on, offering its advice on difficult matters when asked. It was composed, as we have seen, of outstanding citizens in oil and gas or allied industries. It now became a permanent committee with an ever-changing membership, its members selected by each new president. Near the end of his term Sandlin would initiate a by-laws change that would make a retiring president chairman of this select committee in order to maintain some continuity in its membership.

Sandlin, as has been noted, already had filled two vacancies on the committee with appointments of John Mecom and Michel T. Halbouty. Now he persuaded George Brown, Donald Connelly, John Flanagan, William Hendrickson, John T. Jones, George Strake, and Harry C. Webb (the Club gourmet) to keep their seats; to round out the committee he appointed Ben C. Belt, vice-president of Gulf Oil Corporation; J. R. (Butch) Butler, president of J. R. Butler and Company (oil); Leon Jaworski, attorney; Earle W. Johnson, president of General Geophysical Company; Patrick Rutherford, oil operator; George Sawtelle, president of Kirby Petroleum Corporation; and one of the founders, Wilbur Ginther.

Ginther had shepherded Sandlin into the Club. Sandlin was enough of a romantic to want a link with the Club's beginning on this committee. And he valued Ginther's enthusiasm, and his judgment.

Belt of Gulf meant something else to Sandlin. He had joined the Club before its opening, the first executive of a true major company to brighten the Club's door. Gulf was an international giant with prolific foreign production, particularly in Venezuela and Kuwait. Had Belt joined the Club for no reason but to enjoy the gourmet meals, it would have been regarded as a *coup* by the membership. But Belt was a hardworking civic leader, and it is likely that he entered the Club with intent to serve it. However, if the members hoped for a parade of major company executives to fall in line behind Belt, they were disappointed. No parade formed, no bugles blew. (Leonard McCollum, a staunch Club member, had not yet taken his Continental Oil Company into Africa.) Certainly, Sandlin appointed Belt to the committee because he wanted the benefit of his advice. But he also was hopeful that Belt's presence in this prestigious group would prompt the Club members to select him as a director in the near future. "He would make a strong director," Sandlin confided to Ginther, "and his being a director could bring more like him into the Club."

Leon Jaworski was Hugh Q. Buck's colleague at Fulbright, Crooker, Freeman, Bates, and Jaworski. He had gained national attention as a prosecutor at the Nazi war crimes trials after World War II. But even before the war he had established a reputation as a superior lawyer, and some of the cases in which he had been involved were based in oil. In one noteworthy case he had represented Glenn McCarthy, the wildcatter and builder of the Shamrock Hotel. A McCarthy well, which blew in wild and sprayed the countryside with oil and gas and fire residue, was alleged by landowners to have rendered unproductive for all time the acreage around it. Jaworski had delayed the trial until the growing season. Then he showed the jury beautiful color pictures of budding trees and greening grass in the area with cattle grazing contentedly on nature's bounty. Jaworski had no ambition to serve the Club as a director or officer, but he was willing to accept an advisor's role— and Sandlin wanted him on those terms. (McCarthy had been a charter member of the Club, but he had resigned after the Shamrock opening and before the opening of the Petroleum Club.)

Sandlin's tenure was as smooth as he was, though some of its success could be attributed to the momentum his rough-hewn predecessor had generated. Sandlin's goal was to have a club with stability, with character. He loved a free exchange of ideas and knowledge. He wanted to create a homelike atmosphere. (This had as much to do with a House Rule against tipping as the announced desire to prevent a loud-mouthed drunk from demanding superior service while waving a handful of bills at a waiter. Incidentally, Hugh Q. Buck, House Committee chairman, would report to the membership at the end of Sandlin's term that "there have been no drunks in the Club," and the remark drew only scattered laughter.)

At the membership meeting when his term was about to expire, Sandlin had praise for the work of the various committees. The

Library Committee, for example, had laid down the foundation for a worthy collection of reference books, and George Sawtelle of the Advisory Planning Committee had donated to the library a complete set of American Association of Petroleum Geologists Reports.

The Publications Committee, under the leadership of George Barrow, had published a new membership roster, a copy of the by-laws, and had changed the design and style of Club bulletins.

And the Club had been fairly throbbing with excitement in contrast to the dull evenings decried by Leonard McCollum. The Entertainment Committee during the year had arranged for 39 club dances with a total attendance of 6,400 members and their guests. There had been five cocktail dances and buffet suppers preceding football games, with 837 attending. There were two fashion shows, attended by 431 members and their wives, and another show was being planned. And there had been two gin rummy tournaments with 151 members participating.

One of the style shows prompted the first change in the Club's furnishings. It was held in the beautiful Celestial Room, the main cocktail lounge. And Oleg Cassini, the famous designer, was on hand to monitor the affair. He was everywhere, bowing here, kissing a hand there, charming the ladies with his wit and continental suavity. Then he and Halbouty, the Entertainment Chairman, saw something that turned Halbouty's face red with embarassment.

"My good friend," Cassini whispered in Halbouty's flushing ear, "This event is being staged *not* to show off the *under*garments of the guests but the *outer* garments of the models."

The exquisite, comfortable chairs and couches in the Celestial Room were so low that there simply was no way that a woman could sit in them properly.

"Indeed," Cassini whispered to Halbouty, "there is one lady here who has *no* undergarments to display!"

A few members' wives already had complained about the chairs and couches, but Cassini's observations and Halbouty's embarrassment brought about a rapid transformation of the room's decor. Upshaw, who had heard the original complaints, was grateful for the directors' quick reaction.

At this membership meeting—April 20, 1954—Secretary W. O. Bartle told the membership there were 815 resident members in good standing with 85 vacancies. There were, Bartle reported, 418 producers, refiners, and transporters; 61 professional consultants; 38 drilling contractors; 50 service company executives/representatives; 19 lease brokers; 91 equipment supply men; 19 royalty owners; 24 oil and gas lawyers; 19 oil and gas bankers; 76 miscellaneous. There were, he added, 518 non-resident members in good standing with 232 vacancies.

The Club was out of debt, according to Hugh Q. Buck, treasurer. There were adequate funds on hand, and the Club was operating at the "break-even" point. "Our financial status is on an even keel," Buck said.

There was a sad note, however. Fifteen resident members and an identical number of non-resident members had died since the Club's opening. Sorrow at their passing was expressed in a resolution.

Four new directors were elected to serve three-year terms: Ben Belt (as Sandlin had hoped), Donald L. Connelly, Wilbur Ginther (the founding father), and Michel T. Halbouty. They would replace Sandlin, McCollum, David Bintliff, and Virgil Brill.

At the next meeting of the board of directors—the annual organizational meeting—Hugh Q. Buck, the stalwart, became the Club's fourth president, elected by acclamation. Ralph Johnston and E. O. Buck were elected vice-presidents; Ginther was elected secretary; W. J. Goldston was named treasurer, and Halbouty was elected assistant secretary and treasurer.

As Sandlin handed over the reins to Hugh Q. Buck, the outpouring of affection for him by his colleagues was restrained but touching. They put their sentiments in a resolution which said:

> "RESOLVED, that Marlin E. Sandlin, President of the Club for the twelve-month period just ending, has with full devotion to the interest of the Club wisely and energetically executed his duties as President. The efforts that he has extended through this year are but an extension of the real and continuing contributions by him through the years since the organization of the Club, and the Directors feel that they express the genuine sentiment of the membership of the Club in extending to him their heartfelt and sincere thanks and appreciation."

Under the change in the by-laws, Sandlin became chairman of the Advisory Planning Committee.

The 50 service company representatives and the 91 equipment supply men among the members were integral parts of the Club—and of the oil and gas industry. These were the men and companies that constructed the refineries and laid the pipelines. They manufactured the equipment and supplies for wells to be drilled. They provided the techniques that assured that wells were drilled accurately and with a remarkable degree of safety. They peddled their wares and expertise to the majors and independents alike, and there were few oil patch mysteries they could not solve.

Some of the products were born of necessity. The great drilling mud companies like Baroid, Magcobar, and Milchem were the result of Curtis Hamill's ingenuity during the drilling of the Spindletop Gusher in 1901. Curtis Hamill and his two brothers, Jim and Al, had drilled down to 640 feet when they hit a sand they could not penetrate. In 1957, when he was in his eighties, Curtis Hamill told this story to an interviewer:

> "We would lose all the water from our pumps in this sand. Our pipe would stick. Then, one time when we pulled our pipe out, our water supply was gone because the jet in the water well had failed to work. On top of that, we were out of wood for our boiler. We really were a disgusted crew of men . . .
>
> "A Reverend Chaney owned the teams we used in cleaning out the pits. The teams arrived about one o'clock and we got a big pile of sand out of the pits. 'Where in the world did you get all that sand?' Reverend Chaney asked us. I told him we had hit a strata of coarse water sand, and that we were losing all the water we were pumping into the well in that sand. 'I believe that if we could get some muddy water we could stay on top of that sand. We could keep our pump running and get through that sand.'
>
> "Reverend Chaney said that when it rained his corral got deep in mud because the stock tromped around in it. He said his corral was the same kind of clay as our pits . . .
>
> "The next morning we put the cattle in the pit and started putting water in it. We kept the cattle in the pit four hours and they really

tromped the bottom into a muddy slush. Then we put a hose from our drill pumps out to this pit and fastened a piece of one-inch pipe about eight feet long in the end of the hose. This made a good nozzle . . .

"The next morning we mixed the mud some more, then started drilling. At first it didn't do any good. We lost about half of our pit of mud. Then it gradually quit losing and by night we were completely through that formation of bad sand and had drilled into a shale or gumbo formation . . .

"As far as we knew, this was the first mud ever made for the drilling of oil wells. Today we couldn't drill deep wells without the use of mud. . . ."

The "muds" the companies produced were scientifically compounded mixtures of chemicals, ores, and a variety of clays designed to perform different functions under different circumstances. Their representatives were welcomed to the Club.

One early Club member was Sidney Adger of Milchem. Standing six feet, four inches, he had been a star athlete at Louisiana State University. Later he had been regarded as the friendliest—and handsomest—captain in the Pan American Airlines fleet. In the oil patch he was considered a super salesman. He had performed nobly as a member of Halbouty's Entertainment Committee during Sandlin's presidency, and Hugh Q. Buck quickly named him chairman of the committee when he assumed the presidential mantle. (Later Adger would serve the Club as director and officer.)

It has been related how the Cameron Iron Works sprouted from the design for a blowout preventer James S. (Mr. Jim) Abercrombie drew in the dust in the 1920s. Abercrombie never became a Club member, but Cameron executives joined, and so did executives of Abercrombie's oil company.

Even before Abercrombie conceived the blowout preventer, an eccentric wildcatter, Howard Hughes, Sr., set the foundation for another great oil-tool industry in 1908 when he grasped the idea for a tri-cone drilling bit which would revolutionize rotary drilling. In a machine shop at Sour Lake field near Beaumont, Hughes saw a three-wheeled emery wheel with the outer wheels turning in one direction and the inner wheel moving in the opposite direction.

In 1909 Hughes Tool Company was founded in Houston to manufacture the first rotary rock bits equipped with cutters that made it possible to drill rock formations. Ralph Neuhaus, a Hughes' vice-president, was an early Club member and was selected by Bob Smith to serve on the Club's first Advisory Planning Commission. (Later he would move to Cameron Iron Works.) Other Hughes Tool Company men would serve the Club in various capacities and one of them, in 1982, would become the Club's president.

Also on that first Advisory Planning Committee was Pierre Schlumberger, president of Schlumberger Well Surveying Corporation. He was son and nephew, respectively, of Marcel and Conrad Schlumberger, French inventors of an electrical well-logging method that measured the depth and physical properties of the various formations penetrated by the drill. The brothers made the first electrical measurements in a borehole in France's Pechelbronn field in 1927. In 1929, the unique tool was used successfully in the Maracaibo Basin of Venezuela, and in the 1930s, despite initial hostility from geologists, came into use in oil provinces around the world. Houston became a Schlumberger manufacturing, engineering, and research center. Other Schlumberger executives who were early Club members were Robert Rieke and William J. Gillingham. Both would become directors.

Early day Schlumberger technicians, a tough, intrepid breed, followed the drill through swamps and jungles, packing their crude equipment on mule-back when necessary. The equipment would become more sophisticated with each passing decade, and in 1982 Gillingham would note that "the modern generation of logging methods is as vital to the exploration and production of oil and gas as the X-ray is to the practice of medicine. . . ."

Maurice A. Rutis, called the "Pipe Man," played a vital role in the Club's history. He was an executive with the Oilfield Salvage Company and a charter member of the Club. Rutis was a jolly salesman of drill pipe and other equipment and, in the semi-dry state of Texas, he bowed to no man in the procurement of alcoholic beverages. It was natural, then, that he became the Club's provider of refreshments to suit every taste.

In those days, whiskey-by-the-drink and the open saloon were Chamber of Commerce dreams. A man either bought a jug of spirits and drank it, or he became a member of a "private club" where he could saunter up to a bar and order a martini or any other mixed drink. Rutis operated as "M. A. Rutis et al," and no Club member ever went thirsty . . . or worried about a lawman tapping him on the shoulder. So well did "M. A. Rutis et al" conduct business with liquor dealers that Club members always had drinks on the house at annual meetings and other affairs.

And a new kind of oil business was a-borning and providing new members for the Club. As early as 1896, wells were drilled off piers at Summerland, California, and in 1910 Gulf Oil Corporation leased 8,000 acres of the floor of Ferry Lake in Louisiana with plans to initiate overwater drilling. But the true pioneers of offshore drilling were the men who found and exploited the vast treasure trove beneath the waters of Venezuela's Lake Maracaibo in the 1920s and 1930s. The devices and techniques they developed were the prototypes for the sophisticated vessels and platforms that later allowed oilmen to conquer the world's seas and oceans.

In 1938 the first producing well in the Gulf of Mexico was brought in by Superior Oil Company and Pure Oil Company off the Louisiana coast, and in 1941 the first well was drilled off the Texas coast, at Sabine Pass. Great interest in the Gulf was generated in 1946 when Continental Oil Company, Atlantic Refining Company, Tide Water Oil Company, and Cities Service Company formed the CATC Group to test the Gulf. The following year Kerr-McGee Oil Industries, Inc. drilled the first Gulf well out of sight of land—12 miles offshore from Terrebonne Parish, Louisiana.

By the mid-1950s, J. R. McDermott & Company, Inc. of New Orleans had built a fabricating yard solely for the assembly of offshore structures. The Gulf was becoming one of the "hottest" of oil provinces. New companies were formed to create drillships and more and more exotic drilling platforms, and other companies sprung up to service these creations.

Houston, as the oil capital of the world, profited from the new kind of oil business, and the Petroleum Club of Houston was strengthened by the membership of the men who ran it.

In addition to appointing Sidney Adger to head the Entertainment Committee, Buck named Rutis as chairman of the House
Committee, Sam A. Merrill chairman of the Finance Committee,
J. S. Gissel chairman of the Food Committee, G. Flint Sawtelle
chairman of the Library Committee, and George T. Barrow continued as chairman of the Publications Committee. Merrill was
an accountant; Gissel was president of a barge line company, and
Sawtelle worked with his father, George Sawtelle, president of
Kirby Oil and Gas Company.

During the Sandlin administration, Leonard McCollum had
suggested that the Club needed young blood as well as new blood.
He had offered up the possibility of opening the door to younger
men about town on a specialized basis. Now Buck took the Club a
step farther in that direction. According to the minutes,

> "A new class of membership composed of younger businessmen
> would be offered for sale at a price low enough to be attractive,
> and also to induce employers of young executives to make the
> capital investment in the cost of the membership. Such a class
> would probably not carry voting rights, thus properly distinguish-
> ing the class from full membership and justifying the sale at a
> lower price. . . ."

The new class of membership would not be created during
Buck's presidency, but he strongly recommended that the succeeding administration take action on the matter. "The holder of such
a membership," Buck said, "would probably have a right, upon
passing the age of 35 years, to convert his membership into a voting membership." And Bob Smith commented that such a pool of
younger members would "constitute a source of strength for the
Club."

Smith and Don Connelly also recommended that certain members of the board and the advisory Committee get in touch with
former members who had resigned from the Club. "If it turns out
that they're the kind of persons we want in the Club, then we
should try to get them back," Smith said. The directors agreed
that the idea was a good one.

A committee composed of Connelly, E. O. Buck, and Vernon
Frost was appointed to investigate the repurchase of memberships
of deceased, incapacitated, and resigned members. The commit-

tee recommended that the Club, within its financial capability, re-purchase memberships of deceased persons, upon request of the family or estate; of incapacitated persons found by the Board to be in real need of the sale of their memberships; and of those persons who had moved or been transferred to other cities.

It was a compassionate move, but it also made business sense, the committee members pointed out. They recommended "that as additional memberships are sold, the Club transfer to such members the certificates thus purchased, and that such certificates first be exhausted prior to transferring the certificates of any other member. . . ." The Board voted to act on the recommendation.

Members had been complaining about parking facilities since opening day, and Upshaw had noted a slight falling-off of business which he attributed to parking problems. During an "interesting" discussion about these matters, the first whiff of discontent with the Rice roof floated over the Board meeting room. According to the minutes,

> "The matter of future planning incident to the continuation of the finest possible quarters for the Club was discussed. It was decided that this Board would recommend to the incoming administration that a Special Committee be formed for the purpose of appraising the needs of the Club and recommending plans for improvements or changes for the future. . . ."

The terms of Hugh Q. Buck, E. O. Buck, Vernon Frost, and Ralph Johnston expired April 19, 1955. Hugh Q. Buck was a charter member, Johnston and E. O. Buck had been members since 1948, and Frost, of course, was one of the eleven founding fathers. Their departure from the Board left only two members of the old guard as directors—Ginther and Smith.

The four retiring board members were given a rising vote of appreciation at the annual membership meeting in the Club's Main Dining Room. It was a warm tribute, but the moment held a special charm for Frost because of its contrast with the oftime joyless days of the Club's childhood. Looking out at the happy crowd in the splendid hall he could not help but think of the eleven coffee drinkers huddling in the Esperson Drugstore while they talked of a club that would be the finest in the land.

Elected to succeed the four were Sidney Adger, Colonel W. B. Bates, B. G. Martin, and Randolph Yost. Adger, as noted, had been chairman of the Entertainment Committee under Buck. Colonel Bates was a banker and partner in Buck's law firm, Fulbright, Crooker, Freeman, Bates, & Jaworski. Martin was president of San Jacinto Petroleum Corporation, and Yost was vice-president and division manager for Stanolind Oil and Gas.

The next day, at the annual organization meeting of the board of directors, W. O. (Bill) Bartle was elected president of the club. Halbouty was elected first vice-president, Ginther second vice-president, Adger secretary, Bates treasurer, and Yost assistant secretary-treasurer.

Bartle named Martin chairman of the House Committee; E. O. Buck chairman of the Finance Committee; Walter G. Sterling, an early member and vice-president of Royalty Properties, chairman of the Library Committee; George C. Hardin, Jr., a geologist who worked with Michel Halbouty, chairman of the Entertainment Committee; R. G. Rice, executive vice-president of Tennessee Gas Transmission Company, chairman of the Food Committee; Marion L. Martin, an insurance executive, chairman of the Publications Committee.

And in accordance with the recommendation from the previous board, Bartle appointed a Special Committee with Bob Smith as chairman. Smith was joined by Sandlin, Hugh Q. Buck, Bintliff, W. A. (Bill) Kirkland, and S. Marcus Greer. Kirkland was president of the First National Bank and Greer was vice-chairman of the board of City National Bank.

The committee's first assignment, Bartle said, "would be to supplement the work previously done with a continued study of the Club's parking problem." If it appeared that Bartle had loaded his shotgun with buckshot to kill a mouse, it was not the case. Parking was an aggravating problem.

The Rice Hotel Garage was a block and a half away from the hotel proper. It was sadly understaffed. Waiting for one's car to be delivered to the hotel door could destroy the memory of vintage wine and *Le Filet d'Agneau aux Fines Herbes*. To walk to the garage and still wait for one's car was tantamount to waiting for a Supreme Court decision.

The garage manager had answered complaints with a shrug that said it wasn't his fault. Ben Orr, who was in charge of the entire Rice Hotel operation—and who had parried Upshaw's every thrust during the food-service contretemps—had been as unyielding as Strangler Lewis to appeals for more parking attendants.

Recognizing the virtue of a Special Committee, Bartle had staffed it with heavyweights in hope of realizing an immediate gain.

Bill Bartle had never been accused of hesitancy. He had been elected to the Board to serve with Marlin Sandlin, and had been elected secretary by his fellow directors. At the first board meeting, Sandlin had asked if any of the new directors had any observations to make on Club operations. It is likely that Sandlin's question was rhetorical, but Bartle stood up and, in rapid order, listed several areas where he thought changes were in order. Such forthrightness in a freshman director could have been abrasive, but Bartle had brought into the open problems that obviously had been disturbing his seniors for some time but had not been discussed. His boldness cleared the way for solution of the problems.

Bartle was 58 when he assumed the presidency. His formal education had been limited—he had attended Rice Institute for three months—but he was a man of great intellectual curiosity and wide-ranging interests. He was the possessor of a magnificent collection of American gold coins; his passion was the assembling of an entire set of $20 gold pieces struck from a design by the renowned Irish-American sculptor Augustus Saint-Gaudens. Commissioned by President Thodore Roosevelt, the first Saint-Gaudens were minted in 1907, and a limited quantity was minted each succeeding year until 1933 when another Roosevelt ended the coinage of gold.

Bartle delighted in showing the Saint-Gaudens to his closest friends like Walter Sterling, though at the time the set was incomplete. He was lacking one minted in Denver in 1927. Only a handful were know to be extant. It would take years, a world-wide search and considerable expenditures before Bartle would acquire one and round out his prized collection.

Bartle had entered the oil business as an accountant for Wynn Crosby Drilling Company. He had learned accountancy through a correspondence course. While at Crosby Drilling he met Sid Brewster, the Crosby field boss, and in the mid-1930s they set up Brewster & Bartle Drilling Company in Houston.

The company was successful, and after World War II the partners pioneered inland barge drillings. Their barges were busy in marshes and rivers and the shallow waters of Galveston Bay. Brewster & Bartle became the largest company in that business.

Bartle, as president, was noted for decisive action. At the annual membership meeting preceding Bartle's election, Logan Bagby, Jr., an early member, had suggested that a study be made of the wage structure for Club employees, "having in mind that there may be some inequities." At the first board meeting after Bartle's election, the House Committee presented such a study to Bartle and the Board, plus recommendations for a wage system providing periodic increases as an incentive for long-term employment. The system was approved by the Board, and Upshaw was instructed to utilize it as quickly as possible.

The House Committee had another recommendation, one that severed a connection with the past. For the Club's opening, the noted photographer Paul Linwood Gittings had made portraits of the Club's officers, directors, members of the Advisory Planning Committee, members of the House Committee, John Staub, Edward J. Perrault, Jr., and Seymour Fogel.

Gittings had told reporters that it took him eight months to complete the photographs. They were magnificent, and they hung along the brick wall of the Cocktail Corridor leading from the marble-floored entrance area and dining room to the west wing. The photographs were so striking that it was well-nigh impossible for a newcomer to pass them with only a glance.

Now, in June of 1955, the House Committee recommended "that the pictures of the artist (Fogel), architect (Staub), and decorator (Perrault, who preferred 'designer'), be returned to Mr. Gittings in order to make room for adding the picture of the newly-elected president each year." The recommendation was unanimously approved. Perrault, for one, was not disturbed by the action for he had become a member of the Club and was an active committeeman.

If the Board's action appeared a bit cold to some members, Bartle had a sympathetic ear when others discussed the status of widows of deceased members with him. The Board, and later the membership, decided that "the membership stock of any deceased resident member may be held by the widow of such deceased member so long as she desires to use the facilities of the Club. The existence of such outstanding share as held by the widow shall not be computed in arriving at the overall limitation on the number of permissive active members as may be established by the Club from time to time. The widow shall, following the date of the death of the deceased member, be relieved of the payment of monthly dues and any and all assessments other than for the payment for food and other services extended to her."

While the Special Committee was studying the parking problem, Upshaw was noting that the decline in business was becoming more pronounced . . . and that parking alone could not be blamed. For one thing, the Houston Club had moved into bright new quarters, and was attracting Petroleum Club members for lunch. It was plain that Upshaw regarded this defection as temporary, that the Petroleum Clubbers would return to their own tables once the Houston Club novelty wore off.

No, he was concerned because the decline appeared to have been noticed in other clubs in the city. And to make things worse, the Hilton Hotel Company had taken over the Shamrock, had renamed the Cork Club the International Club, and was spending $25,000 to $30,000 a month on topline entertainment.

Despite this gloomy outlook, or perhaps because of it, Upshaw proposed that the Club kitchen be renovated and new equipment purchased. The directors, not to be outdone, voted approval of the project. When it was completed during the following presidential term, it cost $53,377.

The Special Committee brought in several alternatives to the Rice Hotel Garage parking for consideration, but none was really attractive. Finally the committee suggested that Bartle, as president, approach Fred Heyne, Sr. and try to persuade him to increase the parking staff. The personal approach might get the job done, David Bintliff said.

Bartle did go to see Heyne, and he was still waiting for an answer to his proposal when his presidential term expired. (A small parking area and driveway was eventually constructed at the rear of the hotel, allaying the parking problem to some degree, but still the complaints came in to Upshaw and the Board.)

Bartle's term as director expired with his presidency, as did the terms of W. J. Goldston and Bob Smith. Bartle was accorded the by-now customary laudatory resolution by his fellow directors, and they also took note of Smith's eight years of service, four of which he had spent as president. Goldston also was commended for his years of service to the club.

At the annual membership meeting the members elected A. L. Selig, George C. Hardin, Jr., and Logan Bagby, Jr. to the Board.

According to the minutes, "C. S. Birch rose from the floor to thank Mr. Bartle for his splendid service to the Club as a Director and as its President. He said he felt sure he was speaking the sentiments of the membership as a whole. His remarks were approved by a round of applause by the members present."

Selig and Hardin had served the Club as committee chairmen. It will be recalled that Bagby, an early member and worker for the Club, had recommended the employees' wage study at the previous annual meeting. This recommendation was only one of several constructive studies he had initiated.

At the Board's annual organization meeting of April 18, 1956, Michel T. Halbouty was elected president, Ginther was named first vice-president, Randolph Yost second vice-president, Bagby secretary, Selig treasurer, and B. G. Martin assistant secretary-treasurer.

By now the Club had settled down. It ran smoothly, as each administration drew strength from its predecessor. Its members were proud that they belonged to it, and saw their membership cards as badges of distinction. They were prone to admire their officers and directors, and thus were judicious in the selection process.

And the Club had earned an enviable reputation in the city as a whole. It was regarded generally as an asset to the community.

But the Petroleum Club of Houston basically was still a club for independents and their business allies. Many members, no doubt, were content for it to remain that way. But for others, particularly some of the early members, lack of whole-hearted acceptance by the majors was a constant pricking of their pride.

CHAPTER SEVEN

DEMAND FOR OIL in the post-war world surprised even the most optimistic wildcatter. The American independents drilled wells and more wells, but their share of the ever-growing U.S. market remained the same. Their increase in production to supply the mounting demand was more than matched by the importation of foreign oil. They were still "prorated," with their wells limited in both production and days of operation.

Then, in 1956—the year Bill Bartle handed over the Petroleum Club reins to Mike Halbouty—Gamal Abdel Nasser, the Egyptian premier, nationalized the Suez Canal, the great waterway through which was funnelled Middle East oil to western Europe. Britain, France, and Israel attacked Egypt, and British and French forces occupied positions along part of the Canal. The small lumbering oil tankers of the period were re-routed around the Horn of Africa.

Oil-dependent Europeans were threatened by a genuine "energy crisis." They were saved by an "oil lift" from the U.S. and Venezuela. American producers were allowed to open their valves, and the majors ferried the oil to their European outlets. The valves remained open until the international dispute was settled and the small tankers resumed their journey through the Canal.

It had been a heady time for American producers—selling all the crude they could produce for more than $3 per barrel—and they were more than reluctant to return to normalcy. Normalcy meant 8 producing days per month, with each well producing no more than 13 barrels per day. Normalcy, the more vocal among them complained, was "sucking hind teat to imported oil." Imports, they argued, were not "supplementing" domestic production, as the majors insisted, but "supplanting" it.

Independents—and some outside the oil industry—also saw the imports as a threat to national security. Because of limited production there was little incentive to risk fortunes in unproven but promising areas for a limited return, they said, and the pool of risk capital the independents relied on was, of consequence, drying up. The true wildcatter was a vanishing breed, they said, and offered statistics to support the claim. For the majors, with great overseas production, there was even less incentive to hunt new reserves in the U.S., they contended. Therefore, they argued, the imported oil was holding back the search for new reserves the U.S. would need in a time of international crisis.

The oil industry had two great friends in Washington, Senator Lyndon Johnson and Sam Rayburn, Speaker of the House, both Texans and both Democrats. While they protected the industry as a whole against all foes, they were aware that independents centered in their home state were more vital to their careers than were the majors. They prevailed on the Republican President, Dwight Eisenhower, to establish a "Special Cabinet Committee" to determine if crude was being imported in such quantities as to threaten national security.

Eisenhower was moved to institute "voluntary controls" on oil imports, and later "mandatory controls." Both caused great confusion, but neither had a discernable impact on oil imports. The independents then began lobbying for enactment by the Congress for laws establishing import restrictions. The movement died when many leading independents spoke out against it. Mandatory controls, even if they didn't work, could at least be amended, revised, or even rescinded but, as one independent declared, "Only an act of God can change an act of Congress as far as the oil business is concerned."

So the majors-independents relationship basically remained unchanged in the United States.

On another front the independents suffered a major disappointment. Eisenhower had pronounced support of legislation to ease federal control over the natural gas industry, and shortly thereafter insisted that federal gas regulations be repealed. The Congress responded with a repeal bill. In the meantime, it was alleged that a representative of Superior Oil Company had tried to bribe a Midwest senator to vote favorably on the bill. Eisenhower, in high dudgeon, vetoed the bill he had requested while publicly castigating the entire industry for what he called its "arrogance." Many thought he offered a poor reason to veto a bill that ostensibly had been drawn to benefit the American public. Cynics figured that Ike's advisors had convinced him that the public favored maintenance of the regulations—and that more people bought natural gas than produced it.

So natural gas legislation went on a back burner.

Strong, aggressive independents, however, made inroads in the majors' foreign preserves. The first was made in the Neutral Zone, a square of swirling sand between Kuwait and Saudi Arabia. Led by Ralph K. Davies, a former Standard of California vice-president, ten American independent companies and individuals dealt with the Sheik of Kuwait for a concession on Kuwait's half of the Neutral Zone.

The partners were J. S. (Mr. Jim) Abercrombie of Houston; Ashland Oil and Refining Company of Ashland, Kentucky; Deep Rock Oil Company of Chicago; Globe Oil and Refining Company and Lairio Oil and Gas Company, both of Wichita, Kansas; Hancock Oil Company of Long Beach, California; Phillips Petroleum Company of Bartlesville, Oklahoma; Signal Oil and Gas Company of Los Angeles; Sunray Oil Corporation of Tulsa, Oklahoma, and Davies himself.

Shortly thereafter, a man with the wildcatting instincts of Columbus Marion (Dad) Joiner and the acquisitive instincts of John D. Rockefeller obtained a concession on Saudi Arabia's half of the Neutral Zone. His name was J. Paul Getty, and he was destined to play a major role in world oil history, the first individual to invade a region staked out by the giants. He had inherited wealth, but like some others in that category, Getty was determined to show that he could have made it had he been left a foundling on a peasant's doorstep. He obtained his concession under the name of the Pacific Western Oil Corporation, Los Angeles. His name would be associated with other companies over the years, but his name would always be predominant.

Davies' group was called the American Independent Oil Company (Aminoil). With Aminoil as operator, the two companies—after three dry holes and $30 million—brought in Wafra #4, and by the end of 1956 had 59 wells producing from the same rich sand that lay beneath Kuwait's Burgan field.

Getty and Aminoil also pushed their way into Iran, as did several other independents. It will be recalled that in 1951 revolutionaries overthrew the Shah and nationalized the Iranian oil industry during a dispute with British Petroleum, the sole producer of the country's vast petroleum reserves. A successful counter-revolt, fomented with the help of American and British agents, restored the Shah to power. Under an agreement born more of politics than economics, Iran retained ownership of the oilfields, and a consortium was formed to produce and buy the oil.

The consortium originally was made up of British Petroleum, Royal Dutch-Shell, Mobil, Standard of Jersey, Texaco, Gulf, Standard of California, and the French-owned Compagnie Francaise de Petrole (CFP).

But Getty and the Aminoil group, fresh from their success in the Neutral Zone, demanded entrance into Iran, and they banged on the consortium door until it was opened. Parading behind them were Standard of Ohio, Atlantic and Richfield (they would merge), Signal and Hancock (they also would merge), and San Jacinto (which would be bought by Conoco). The independents were herded together under the banner Iricon, and given five percent of the action. British Petroleum held 40 percent, Royal Dutch-Shell 14 percent, CFP 6 percent, and the 5 American-based giants held 7 percent each.

In the Neutral Zone it had cost the independents four years and $30 million to uncover a bonanza. In Iran, where much of the searching already had been done, obtaining that precious five percent was like winning a million dollar lottery at the expense of a stamp and the licking of an envelope.

The independents were eager and ready when Libya decided to open its doors to oilmen in 1955. And Libya was eager and ready for the independents. Fifty-one concessions were granted to seventeen companies. The majors were not excluded. Indeed, a Jersey company, Esso Standard Libya, brought in the first well. But the independents obtained many choice concessions, and when volume production was achieved across the country, independents had generated half of it. Conoco, Marathon (Ohio Oil Company), and Amerada, under the banner Oasis, were particularly active, and Conoco's continuing success abroad would make it a semigiant.

The Italians and Japanese, both non-producers but heavy oil consumers, made deals for concessions with Arab and African nations, and the Soviets, with surplus crude from newly-found fields, began cutting into some of the majors' markets.

And back at home the innards of Texas and other oil states were groaning with unproduced crude.

Another great glut was cresting, and everyone who owned an ox cried out that it was being gored.

Had the club's sixth president, Michel T. Halbouty, lived in the 16th century, he likely would have been a swashbuckling pirate, combing the seas for gold-laden galleons while ducking hurricanes and the headman's swishing axe. As it was, he was a wildcatter, and he had eaten a pound of defeat for every gallon of victory he had drunk. His failures were as spectacular as his successes. (A fellow oilman was only half-joking when he told Halbouty's biographer in 1979: "Mike's an exaggeration. He's meaner and sweeter than the rest of us. Tougher and gentler. Smarter and more foolish. Stingier and more generous. Whatever word you pick, that son of a bitch is more of it than anybody else!")

He had made his name known early; six weeks after his graduation from Texas A&M University in 1931, the young geologist-petroleum engineer had been credited with locating a rich oil pool beneath the overhang of a mushroom salt dome at High Island, Texas. Over the years, he had found other fields—and had missed some by inches. He had made fortunes and lost them.

Now, in 1956, as he assumed the club presidency, he was skipping in high cotton because he had just discovered the Pheasant field in Matagorda County, Texas, and had found a prolific new reservoir on the southeast flank of the old West Hackberry field in South Louisiana. (Before his term was over, Halbouty would bring in the Fostoria field in Montgomery County, Texas, and the Port Acres field in Jefferson County, Texas.) It is likely, however, that a successful wildcat venture could not have pleased him more than being selected by his peers to preside over the Petroleum Club of Houston.

To help guide his administration Halbouty named Sidney Adger as chairman of the House Committee; R. R. McLachlen chairman of the Entertainment Committee; Herbert M. Seydler chairman of the Finance Committee; Walter G. Sterling chairman of the Food Committee; Charles W. Alcorn chairman of the Library Committee; Willard Gill chairman of the Publications Committee; and to the Special Committee he appointed Hugh Q. Buck, J. R. Butler, Cecil V. Hagen, Alvin S. Moody, and Marlin Sandlin. Buck was named chairman.

Newcomers among the committee chairmen were McLachlen, a Rheem Manufacturing Company executive; Alcorn, vice-president at Falcon-Seaboard Drilling Company; and Gill, a geologist. On the Special Committee, Hagen was a geologist and Moody was chairman of the board of Gulf Coast Drilling & Exploration Company.

There was no mistaking the mission of the Special Committee. It was to look for new quarters. There was no sense of urgency about it, no compelling desire to flee the Rice roof. There were complaints about the present quarters, to be sure. Parking remained a problem. Kitchen space was limited. Because the Club was E-shaped, it was too segmented for some affairs. But these were mere annoyances when stacked against the beauty of the Club, the atmosphere the camaraderie engendered, the marvelous food, the superlative service.

Many far-sighted members, however, had expressed concern to the directors about where the Club would be quartered *after* the Rice lease expired nine years hence. It was not too soon, they said, to begin planning for a new home in a city that was growing like a banyan tree. Nine years would come and go before one knew it. Let's not be unprepared.

Not often voiced was the apprehension that the business district was moving southward as predicted at the Shamrock Hotel's opening in 1949, but it was southward that the committee looked.

Notes written by Hugh Q. Buck after an early committee meeting demonstrate the range of their thinking.

> "It may be possible to arrange with the builders of some new building more centrally located for the ownership of some floors that the Petroleum Club might construct on top of such new building. If the Petroleum Club should now contribute say $200,000 to the equity requirements of the builders, the foundation of the new building could then be constructed to take into consideration the two, three, or more, floors that the Petroleum Club may wish to construct ten years hence.

> "These floors would be separately owned, sometimes referred to as horizontal ownership, and except for elevator service would be completely independent of the base building. The new floors could support their own elevators from the base building roof up. The $200,000 or $250,000, as the case may be, that would presently be paid by the Petroleum Club would have a value to the user between now and the ten-year period at which the Petroleum Club's construction would follow . . . the value of the principal plus the value of the use of the money during that period. The Petroleum Club would thereby purchase in effect the sky rights over any newly constructed base building. . . ."

Buck also advised that discretion be exercised, saying "an effort would be made to avoid disturbing the present lessor-lessee relationship. . . ."

He already had measured a vacant lot at the southeast corner of West Dallas Avenue and Travis Street, and found that it fronted 250 feet on Dallas and 120 feet on Travis. "If these be the true dimensions, the space on each floor would be slightly less than 1,500 square feet. The Petroleum Club would in effect own a city lot at a height of the roof of the base building. No ad valorem taxes would be payable pending the ten years prior to construction. . . ."

He noted that Alvin Moody was going "to make certain investigations regarding the development plans that may now be underway for the construction of a base building at the above mentioned location, and as to whether or not it would be possible for us to deal with the owners thereof. . . ."

(Buck's notes prompt bittersweet recollections of Harris Underwood's memoranda on his studies of the Cotton Hotel, the Ben Milam Hotel, and other possible sites for the Club. Buck could write confidently of advancing $200,000 or $250,000 toward construction because the money was in the club's bank accounts, and surplus at that. During the Club's first two years of existence, there was never more than $3,600 in the kitty, and that was in escrow. There was less than $20,000 in escrow when Bob Smith took over in early 1949, and the Club was $175,000 in debt when it opened in December 1951. Interestingly, the property Buck examined adjoined Foley's Garage, and the garage sat on land previously occupied by the apartment house at 801 Lamar Avenue. Herman Estate Trustees sold the land for hundreds of times more money than they would have received had they leased the old building to the Club.)

So the hunt began, and a calm, deliberate hunt it was. This time the hunters were in a position to pick and choose. This time they found the welcome mat out wherever they went.

The Associate Membership for young executives, conceived by Leonard McCollum in the Sandlin administration and refined through the Hugh Q. Buck and W. O. Bartle administrations, came out of the pipeline at a director's meeting on August 13, 1956. Associate members would be from 21 to 39 years of age. They would pay escalating initiation fees and reduced dues until age 40. "If never having been charged by the House Committee for infraction of the rules of the Club, and had currently maintained his accounts owing to the Club with reasonable promptness at all times, he shall as a matter of right be entitled, upon payment of all fees then due, to all privileges attending the issuance to him of a certificate evidencing a share of stock. . . ."

At that same meeting, Randolph Yost resigned from the Board because his company was transferring him to Tulsa, Oklahoma. Don Connelly nominated J. R. Butler to fill the vacancy created by the resignation. B. G. Martin seconded the motion, and Butler was unanimously elected to serve the remainder of Yost's term.

A kink in the club's membership selection process was smoothed out during the 1956–57 club year. Applicants were required to obtain letters from two recommending sponsors before being considered for membership by a secret Membership Committee appointed by and known only to the Club president. If the Membership Committee deemed the candidate worthy, his name was passed on to the Board. The Board had the candidate's name posted on the club bulletin board for inspection by the membership. If the candidate passed this crucial review, his name went back to the Board for approval of his membership.

This was fine, but Colonel Bates pointed out to the Board that from time to time a director appeared as one of the applicant's two sponsors. "This amounts to automatic approval," Bates said.

Yes, others agreed, and it is hard to say no when an applicant asks that you sponsor him.

"All right," said Bates, "I move that no member of the board of directors shall be a recommending member on applications for membership."

Ginther seconded the motion and it was passed unanimously.

While business of this kind was being handled expeditiously by the Board, the Club was enjoying a busy social season. And the Club also opened its doors to the Texas Mid-Continent Oil and Gas Association, which was convening in Houston. The conventioners ate breakfast at the Club and held a reception as well.

A suggestion by Don Connelly that the club hold a special buffet luncheon stag party during the Christmas season to which members could bring guests evolved into the Annual Fellowship Luncheon. The cheerful mingling at these yearly gatherings appeared to heal old wounds and erase older scars. New friendships were created, too, and ancient ones renewed. (Though supplies of food and drink appeared unlimited, it cost only $5 per person to attend that initial Fellowship Luncheon!)

Prior to the Christmas season, Halbouty received a letter from Edward Marcus of Neiman-Marcus in which Marcus offered to furnish a Christmas shopping service to members. According to Secretary Bagby's minutes, here is how the matter was handled:

> "The idea was to have on display in the foyer during the luncheon hour a selection of gifts, and also a person representing Neiman-Marcus who could make suggestions and take orders, since many members have a very limited time in which to shop and that this service would be of some assistance.

> "After discussing the matter, it was generally agreed that although there could be a beneficial result in such an arrangement, it might set a precedent which may not meet with full approval of the members.

> "Mr. Halbouty stated that in order to reply to Mr. Marcus he would like a motion to be voted on, and he, therefore, made a motion that was seconded by Sidney Adger that Neiman-Marcus's offer of a special Christmas shopping service set up in the Club foyer be accepted. The motion failed to carry, and Mr. Halbouty said he would inform Mr. Marcus. . . ."

Thus the Board, not Halbouty, would say no to Marcus. And, if he so desired, Halbouty could truthfully tell the merchant that he had proposed and voted in favor of the display. It was an ancient ploy, one in general practice in corporation board rooms, but the Club directors over the years apparently felt they had to use it with some regularity.

From time to time the directors had to turn down gifts from outsiders and Club members alike. Take the case of Michael V. Kelly, oil operator and early Club member, who telephoned Halbouty to offer a gift to the Club. Halbouty told him to write him a letter he could present to the directors. Kelly wrote:

> "As discussed in our recent telephone conversation, I am sending this letter in accordance with your request so that you may present to the Board of Directors of the Petroleum Club the idea of using a trophy moose that I killed in Canada last month for the library in the Club. This head is now being mounted in Cody, Wyoming and should arrive in Houston sometime around December 15. I would be most happy for the Petroleum Club to have this trophy if the Directors so desire. Please let me know as soon as you can the decision of the Board."

In the club minutes, Bagby wrote tersely, "The very kind offer of Michael V. Kelly to give a trophy moose head to the Club was discussed and it was decided as a matter of policy to decline the offer."

And Halbouty wrote Kelly:

> "Your request to make available a trophy moose was taken before the Board of Directors today, and your gracious offer was declined because of existing policies that had been set by previous boards of not accepting animal trophies of this kind because of the difference in decor as well as not wanting to establish a precedent for such gifts. The Board of Directors, however, wanted me to take this means of thanking you for your generous thought in behalf of the Club. . . ."

(The proud hunter, a short time later, would be elected to the Board by the membership.)

Secretary Bagby, meanwhile, was responsible for the reversal of an established club policy. Almost from the Club opening, Bagby and various other members had asked succeeding board to consider establishing reciprocal arrangements with certain other clubs throughout the country. Each board had turned aside the requests for what appeared to be a valid reason—the loss of potential non-resident memberships. Why, Hugh Q. Buck had asked on numerous occasions, would a member of the Petroleum Club of New Orleans, for example, buy a non-resident membership in the Petroleum Club of Houston when, if the clubs were reciprocating, he could enjoy the Petroleum Club of Houston for nothing? And, non-resident membership fees and dues were an important part of the Petroleum Club of Houston's annual revenues, he would point out.

Even before the club was completed, Marlin Sandlin had negotiated reciprocal agreements with the Calgary Petroleum Club and the Edmonton Petroleum Club. But both were in Canada and presented no threat to sale of non-resident memberships. It was arrangements with domestic clubs—particularly those in Texas and other oil states—that were consistently turned aside by successive administrations.

But the Club had grown in numbers and wealth, and Bagby's stature in the Club likewise had grown. So a committee, with

Bagby as chairman, was appointed to study the matter. There was no immediate change, but some time later, when Bagby had ascended to the Club presidency, reciprocity was accepted. Eventually, arrangements were made with Capital City Petroleum Club of Jackson, Mississippi; City Club of Baton Rouge; Petroleum Club of Anchorage; Petroleum Club of Fairbanks; Petroleum Club of New Orleans; Petroleum Club of Oklahoma City; Petroleum Club of Tulsa; Singapore Petroleum Club, Ltd.; The Los Angeles Club, The Metropolitan Club, New York; The Mid-America Club, Chicago; Petroleum Club of Denver; Petroleum Club of Los Angeles; Petroleum Club of Wichita; and Westchester Country Club, Rye, New York.

One fine noon as Halbouty strode across the Rice Hotel lobby to the elevator bank he was accosted by a slender man in a tan suit. Standing squarely in front of Halbouty, the man said sternly, "Let me see the palms of your hands, and be quick about it!"

Startled, fearful that he was facing a madman, Halbouty took a backward step. The man smiled. Halbouty exhaled noisily, and grinned. "I'll be damned! It's Lone Wolf Gonzaullas!"

Their memories took them back more than twenty years to a night in the East Texas oilfield, to a vast tent city called Newton Flats, a lively community devoted to sex, pop-skull whiskey, and games of chance. Halbouty was buying a drink for a dark-haired hooker at a pine-slab bar when Texas Ranger Gonzaullus, the only lawman of consequence in the area, had approached them and demanded that Halbouty show his hands, palms up.

Gonzaullus held a theory that if a man's palms were calloused, he was a workingman and presumably honest. If the palms were smooth, the man was incarcerated until Gonzaullus checked him out.

Halbouty, predictably, had refused to show his palms. Gonzaullus, predictably, had drawn a six-shooter and trucked Halbouty off to the new Kilgore jail.

Now, Gonzaullus, who was in Houston on business, was persuaded to join Halbouty for lunch in the Petroleum Club. They were joined by Howard Warren, who had less violent memories of

Gonzaullus and East Texas. They ate lunch, and then the oilmen showed the Ranger around the magnificent club. As they split up to go their separate ways, Gonzaullus smiled and said, "Don't let this place make sissies of you."

Halbouty, Ginther, Connelly, and Belt retired as directors in April 1957. To fill their seats the membership elected Walter Sterling, Ernest D. Brockett, Joseph B. Kennedy, and Howard Warren. Sterling, of course, was an early member and hard-working committeeman. Brockett was a Gulf Oil Corporation vice-president and Kennedy held the same rank with Sinclair Oil and Gas Company. Warren, a founding father, had been the Club's first elected secretary.

At the annual membership meeting, where the election results were announced, Marlin Sandlin called for—and got—a rising vote of thanks for Halbouty's service as president. In turn, Halbouty vowed to continue to work for the Club's best interests.

At the annual organization meeting of April 17, 1957, J. R. (Butch) Butler was elected president, Sidney Adger was elected first vice-president, A. L. Selig was elected second vice-president, and Howard Warren was elected to his old job of secretary. Kennedy of Sinclair was elected treasurer and Walter Sterling was elected assistant secretary-treasurer.

He was born Roscoe Leroy Butler. He detested Leroy and disliked Roscoe, so he told everyone his name was Johnny. It didn't "take." The other kids in Shawtown, Ohio called him Scoe—short for Roscoe. But in 1932, in the East Texas oilfield, he was tagged with a nickname that "took," and he would bear it as long as he lived.

Here is what happened: He was 23, an engineer for the Ohio Oil Company. He and his wife Mary were living in a 9 × 12 shack in a cotton patch near Henderson. Similar shacks were scattered about the patch. One evening a party was being held in the shack occupied by Peter Brainard and his wife "Dirty." Hosts and guests were full of home brew and animal spirits. "Dirty" complained that her ice box pan had rusted out and she pointed to water pooling on the pine-slab floor.

"I'll fix it," Butler said gallantly. He went outside to his truck, returned with a six-shooter, and shot the floor full of holes. "See," he announced triumphantly, "the water's draining off!"

"You butcher!" "Dirty" screamed. She chased Butler out of the shack, shouting after him, "Butcher! Butcher! Butcher!"

For several weeks after that his friends—and others who had heard the story—called him Butcher. Then it was shortened to Butch. Wherever he went thereafter, he was known as Butch Butler.

Butch was a "tough" name, and it fit him. He was stocky and powerful, with marvelous physical coordination, and he was as rough as a woodhauler's butt. In high school he had been an All-State basketball player and had run the 100-yard dash in 10 seconds. At Miami University of Ohio he was an All-Conference running back. Two team-mates would become legendary coaches—Quarterback Paul Brown and Guard Weeb Eubank.

He had worked during the summers for Ohio Oil in the Lima-Findlay field, and upon his graduation the company put him to work in an Illinois refinery for a year before sending him to East Texas. He had studied geology in college. This, plus his summer oilfield experiences qualified him, in that primitive period, as a petroleum engineer. He had become chief petroleum engineer by 1937 when Ohio Oil sent him to the CottonValley field in North Louisiana as district foreman and production superintendent.

In 1941 the companies operating in the field formed the Cotton Valley Operators' Committee with plans to install the first large cycling operation in a unitization program. Butler wanted to run the operation as general manager but his employment by Ohio Oil precluded it. If he quit Ohio Oil, there was no strong reason to believe he could get the general manager's job. He went to see H. L. Hunt, whose influence with other operators in the field outweighed his production. Hunt was impressed with the strapping, assured young man.

"How old are you?" Hunt asked.

Butler said he was 30.

"If a man hasn't got any sense by the time he's 30, he'll never have any," Hunt said. He apparently thought Butler had sense because he said he would help him get the general manager's job. "If you can't hack it, I want to be the first to know, not the last," Hunt said.

Butler got the job, and hacked it. He also got hacked up somewhat when it came time to award a contract for the compressors needed for the cycling. Butler was leaning toward a certain company, and he was discussing the big order with a company official in a cut corn field. With the official was a young giant with a crewcut. He looked as strong as a bear.

"Tell you what," Butler said to the company official. "If that big son of a gun can pin my shoulders, you've got the contract."

It was a lucrative deal, so the company official didn't spend too much time pondering Butler's proposal. "Get him," he told the grinning young giant.

For half an hour Butler and Crewcut wrestled across the corn patch with the short, tough stubble slashing them until they were as bloody as hog killers. Finally the giant pinned Butler's shoulders to the ground. Butler grinned up at the company official through a mask of dirt and blood. "It's yours," he said.

Butler left Cotton Valley field to run a water pressure maintenance project at the Haynesville field, and later moved back to Texas as executive vice-president of the J. S. (Mr. Jim) Abercrombie Company to supervise the unitization of the Old Ocean field.

He went for himself in Houston in 1948 as an independent operator and consultant and, with associates, moved into other areas of the oil business. He was successful. By the time he became the Petroleum Club of Houston's seventh president in 1957, he was organizing Transwestern Pipeline Company to pump natural gas to Southern California.

In the meantime, he had become an outstanding golfer, a strong supporter of Republican Party candidates, and an active member of industry associations. He had joined the Club in February 1949 and had been an influential member. It will be recalled that the Board selected him to fill a seat vacated by Randolph Yost in August 1956. In less than a year as a director, he was elected president.

He was strong-willed but personable, characteristics he shared with his predecessors. He was capable of carrying on their policies and instituting new ones if needed.

During the Halbouty administration Willard Gill and his Publications Committee had commenced work on the Club's first pictorial directory. It was a frustrating job getting some 800 members
to answer requests for photographs and biographical material,
and the job was but 80 percent complete when Halbouty handed
the reins to Butler. Butler wisely retained Gill as Publications
Committee chairman to oversee the directory's completion. In
rounding up the remaining 20 percent of photographs, Gill explained: "The directory is designed for a very useful purpose. It
will serve not only to assist each member in identifying his 800-
plus fellow members, but will include the Club bylaws, house
rules, information on the use of private rooms, and a quantity of
items that will suggest ways in which each member may derive
more pleasure and service from the Club. . . ."

Butler also retained Walter Sterling as chairman of the Food
Committee. The committee had done an outstanding job in supervising the contracting and construction involved in the remodeling of the Club kitchen, a project that was begun during Bill
Bartle's term and completed during Halbouty's. So well did the
committee and Upshaw work together that food service to members was never disturbed.

And Butler also retained Herbert Seydler as chairman of the
Finance Committee. Seydler, it will be recalled, was the first, and
for a while, the only member of the Food Committee. Appointed
by Bob Smith, he had worked with Marlin Sandlin in the negotiations that led to the Club's assumption of food and service from
the Rice Hotel.

F. W. Bell was selected as chairman of the Entertainment Committee and D'Arcy M. Cashin was named chairman of the Library Committee. Bell was manager of Brown & Root's Chemical
Department and Cashin was a geologist and consultant. Selig was
made chairman of the House Committee.

Halbouty automatically became chairman of the Advisory
Planning Committee, but Butler named him to head the Special
Committee. He was joined by Hugh Q. Buck, Alvin Moody,
Marlin Sandlin, and Don Connelly.

Before Butler's administration could get in gear, Ernest Brockett of Gulf resigned from the Board when he was transferred to
Pittsburgh. By a majority vote of the Board, R. R. McLachlen of

Rheem Manufacturing Company, who had served as chairman of the Entertainment Committee under Halbouty, was elected to fill the vacancy.

Brockett was still in the process of leaving town when Kennedy of Sinclair resigned from the Board. He, too, was being transferred. Several nominees for Kennedy's seat were supported, and E. Clyde McGraw won the election. McGraw, president of Transcontinental Pipe Line Corporation, was an early Club member. Kennedy's position as Club treasurer was taken over by Logan Bagby.

A continuing decline in business was in the minds of the directors throughout Butler's term in office. The Eisenhower Administration was struggling with an economic recession, and some of the Club's woes could be attributed to business conditions generally. Not the least of the Club's difficulties was the continuing successful operation of the Shamrock-Hilton's International Club with its big entertainment budget.

But despite the drop in revenue, the directors set about establishing a retirement fund for employees. It was a responsibility they had to assume, Walter Sterling told his fellow directors. He pointed out that some of the employees had served the Club from the beginning. Upshaw had molded the serving corps into the finest in the city, he said, and it would be foolish not to make every reasonable effort to keep it that way. The other board members agreed.

And they refused to cut away at the entertainment budget. Members had become accustomed now to a diversified program of entertainment and activities and extensive use of the Club facilities. Saturday night dinner and buffet dances with top local bands providing the music had become regular affairs. On week nights there was dancing in the Lounge and Dining Room to music from smaller musical groups. Christmas and New Year's dances were anticipated and well attended. Buffets were held before football games with transportation furnished to and from the stadium. Golf tournaments, gin rummy tournaments, and barbecues were of the first order. The Annual Fellowship Luncheon had taken hold. And the Annual Petroleum Club Ball had become a social event of the year. It had been known as the President's Ball in the beginning, but Bill Bartle had insisted on the name change during his administration.

In the early years the directors had actively sought publicity for such events to create a wider interest in the Club and thus entice new members. George Kirksey, it will be recalled, was the Club's first public relations man. He had left these chores behind him to devote his time and energy to help bring major league baseball to Houston. He had been replaced by the firm of Clark, Kemp, and Hazelrigg. James A. Clark was a noted oil historian and *Houston Post* columnist. Club members were the source of many of the fascinating stories in his "Tales of the Oil Country," a weekly *Post* feature. As the Club matured, and the desire for public exposure decreased, Clark and Hal Hazelrigg worked less with the Entertainment Committee and more with the Publications Committees.

Butler had appointed D'Arcy Cashin to head the Library Committee with instructions to consider the Club library as more than a repository for donated volumes. Butler wanted a program of lectures and forums for members, and Cashin assembled it. First on the program was Dr. J. Brian Eby, consulting geologist, who reported on a visit to Russia. He was followed by Dr. Carey Croneis of Rice University, who described the oil potential of the Amazon Basin. Similar lectures and discussions on industry tax problems and other industry concerns became routine functions.

The library was equipped with a projector, and members used it to show films taken on extended trips. And Walter Sterling arranged with Humble Oil and Refining Company to show filmed highlights of Southwest Conference football games, which the company sponsored on radio and television.

The Special Committee investigated several future club sites but found none to its liking. Halbouty, the committee chairman, was questioned about the committee's work at the annual membership meeting of April 15, 1958. He said that the only interesting possibility discovered was a building being planned on a Main Street site. If constructed, the building would be 35 stories, and the committee was studying the possibility of buying the two top floors. He cautioned that the committee was working quietly, that prudence demanded it. By a show of hands, a majority of the members let it be known that they wanted to move southward from the Rice Hotel.

Butler, Adger, Bates, and B. G. Martin departed the Board at this meeting. Elected to their seats were Seydler, H. F. Beardmore, R. R. Dean, and Michael V. Kelly, the moose hunter. Beardmore was a Warren Petroleum vice-president and Dean was president of Bay Petroleum Company and senior vice-president of Tennessee Gas Transmission Company.

The next day, at the Board's organization meeting, Logan Bagby was elected president, A. L. Selig and George Hardin were elected vice-presidents, Kelly was elected secretary, Seydler was elected treasurer, and Beardmore was elected assistant secretary-treasurer.

CHAPTER EIGHT

TWO SHORT LETTERS—written with exasperated affection—
reveal the eighth president's commitment to the Petroleum Club
of Houston. The first, to Club Manager Upshaw, was written one
month after the Club opened. It said:

> "Mr. P. C. Bundy and I took two out-of-town guests to lunch yes-
> terday at the Petroleum Club. As these men were both from Dal-
> las we would have liked to have done a little bragging on our fine
> new Club. Instead we had to apologize for the food and service.
> Any day of the week you can get a good lunch of tasty, well-pre-
> pared food at the Sakowitz Sky Terrace. Why can't we do as well
> at the Petroleum Club?"

The second letter was to Bob Smith. It said:

> "I have today received two advertisements from Paul Linwood
> Gittings, the first sentence of the first paragraph of which reads as
> follows: "Have you seen *my* exhibit at the Petroleum Club?"
> (Emphasis mine)
>
> I was under the impression that these portraits had been pre-
> sented to the club and were the property of the club, and not the
> property of a commercial photographer who is using the Petro-
> leum Club as a setting for an advertising exhibit.
>
> "If these portraits are the property of the club, it is my suggestion
> that they be presented to the individual families of the respective

subjects as a token of the club's appreciation of their efforts on its behalf. If they are the property of Mr. Gittings, I suggest they be returned to him.

"Either course of action will discourage this practice of the exploitation of the club premises for the commercial benefit of an individual."

Bagby's justifiable pique with Gittings demonstrated his devotion to the Club, and his pride in it. From the time he became a member in May 1950 through and beyond his presidency, he was quick to point out what he considered operational deficiencies in the hope they would be erased. Few faults escaped him. His administration would be marked by such attention to detail. He would attempt to tie up loose ends and dust in dark corners.

Bagby retained Willard Gill as chairman of the Publications Committee and Francis W. Bell as chairman of the Entertainment Committee. He named Walter Sterling as chairman of the House Committee, and he appointed three newcomers to chairmanships—Adolphe G. Gueymard, vice-president of First City National Bank, Finance; Ivan G. Burrell, assistant division manager for Ohio Oil Company, Food; William Hurst, petroleum engineer, Library. Halbouty, Hugh Q. Buck, Butler, Alvin Moody, Sandlin, and Bob Smith were named to the Special Committee.

At the organization meeting of April 16, 1958, Bagby called for a vote to spend $2,000 to correct faults in the air conditioning system. And there was some discussion about the employees' retirement plan.

At the May 12 board meeting a committee composed of McGraw, Dean, and Seydler was assigned to give further study to the retirement plan. And there was something else. According to Secretary Kelly's minutes: There was a general discussion on whether or not all of the Club's quarters were being used to the best advantage.

"The fact that the library and card room occupy considerable space and receive little use was brought out. It was mentioned that consideration may be given to the possibility of making alterations in this vicinity (the West Wing) that may provide additional dining room space especially for larger private parties. The President felt that it would be worthwhile for the various Board members, at their leisure, to inspect the West Wing, and present any ideas for making it more useful at the next Board Meeting. As a stimulant to night business, it was suggested that perhaps Bingo would be of interest to the members. It was suggested that the Entertainment Committee give this matter some thought, including an estimate of the cost of equipment, etc."

At the June 9 meeting, on recommendation of the Finance Committee, the Board voted to invest $60,000 of reserve funds in six-month certificates of deposit in a local bank at the best available interest rate. This brought on a discussion of rotating the club's bank accounts because several Club members were connected with various banks. Some time back some of the Club funds had been moved from the National Bank of Commerce to the Bank of the Southwest. Now Walter Sterling proposed that the Club's general account be transferred from National Bank of Commerce, "which has enjoyed the account for a period of some eight to ten years," to the First City National Bank. The motion was carried by unanimous vote. (Jesse Jones had died two years earlier.) There was more discussion about employees' retirement plan, more discussion about Bingo, and a budget of $10,100 for the Entertainment Committee was approved.

At the July 14 meeting there was a sustained discussion of the retirement plan with outside counsel present. Bagby announced that a reciprocal arrangement had been made with the Denver Petroleum Club.

There was more discussion about the retirement plan at the August 11 meeting, as well as discussion about allowing ladies to use the club at noon.

Should ladies be invited to use the club at lunch? It was an old question. At the Club's opening the news media had made much of the fact that ladies were permitted in this masculine domain only after five o'clock, and then in the Main Dining room only.

Wives had nudged husbands. Husbands had entreated board members. The directors gave a little. The appropriate House Rule was amended to read: "All of the Club quarters, except the West Wing, shall be open to ladies on Sundays and after 3:30 p.m. during week days."

But still no lunch time for the ladies. The pressure continued, however, and again the House Rule was amended. It said: "All of the Club quarters, except the West Wing, will be open to ladies on Saturdays and Sundays, and after 2:30 p.m. during week days, and ladies will be admitted to the Men's Grill after 5 p.m. daily."

The more sophisticated early directors—or the more cynical ones—had reasoned that the Club would not be an attractive luncheon spot for ladies once the novelty wore off. Driving into town and finding a parking space eventually would discourage all but the stout of heart. Women, they reasoned, were more comfortable at country clubs with their distractions and more leisurely pace. Certainly ladies were welcome to the Club at dinner time and at parties and other social functions. But not at noon.

But the pressure persisted, and by the late 1950s it was inspired as much by economics as it was by gallantry. The nagging recession had Upshaw and the directors looking in every direction for means of increasing club revenues.

At the September 8 meeting Hal Hazelrigg, the public relations man, was on hand to represent the Publications Committee in a discussion about improving the quality and coverage of the Club's publication "Topics." Improvements would cost money, Hazelrigg said. Some of the increase could be offset by selling advertising space, he said. Selling advertising space had been discussed in other administrations, and the idea had been summarily rejected. Now it was decided to sell advertising space, limiting it to no more than 20 percent of the magazine, with the space being confined to a one-inch strip across the bottom and top of each page. Only the petroleum and allied industries could buy the space. Publication cost could be no more than in the previous year.

(As it turned out, nobody was interested in buying the space for advertising, and the publication subsequently was gently folded.)

It also was decided that clubs in Texas, Oklahoma, and Louisiana not be considered at this time in making reciprocal arrangements, but that clubs elsewhere in the country be considered.

And the employees' retirement plan was pushed ahead another step toward realization.

At the October 13 meeting Upshaw announced that six companies had submitted plans and bids on the retirement plan. He was instructed to give them to Art Bryan of Transcontinental Pipe

Line Corporation who, acting as a consultant, would analyze the programs and make a recommendation to the Board.

At the next meeting, on November 10, it was pointed out that amendments to the bylaws were proposed and acted on at the annual membership meeting where attendance generally was sparse. Would it not be better to permit voting on amendments to be done by mail ballot, thus allowing any changes to represent the desires of a larger share of the membership? Yes, the Board agreed. Hugh Q. Buck was asked to prepare a resolution permitting mail balloting to be presented at the next annual membership meeting.

On hand at the meeting was Art Bryan, consultant, who delivered his studied opinion on the merits of the retirement plans submitted by six companies. The Board voted to award the contract to Tennessee Life Insurance Company.

And, according to Secretary Kelly's minutes, "Once again the subject of allowing ladies to use the Club during lunch hours was introduced." There was talk of removing partitions and installing collapsible doors to provide more space and flexibility, and Upshaw was authorized to ask an architectural firm to make a preliminary study.

Drawings for the proposed alterations were ready for Board perusal at the next meeting on December 8. After some discussion it was decided to carry over the matter until the next board meeting. Meanwhile, a letter would be sent to the membership explaining the plans and asking for opinion.

Bagby was not present at the January 12, 1959 meeting, so A. L. Selig, first vice-president, acted as chairman. He announced that the employees' retirement fund had become effective on January 1 with 100 percent of the eligible employees participating. Bagby, he said, would appoint a committee to administer the plan.

The letter sent out asking the membership for a response to the plans for remodeling the club's West Wing had prompted 455 replies, Selig said, with 376 favoring implementing the plans, 74 against, and 5 straddling the fence. Thereupon, the House Committee was authorized to secure detailed plans and specifications on the job which, if approved by the Board, would be submitted for contractors' bids.

Upshaw brought forward the names of five members whose indebtedness to the Club was in excess of their equity ownership. He said most of the obligations were for dues; repeated attempts to make the members pay up had failed. A motion was made by Herbert Seydler and seconded by Walter Sterling that Section 5 of Article IX of the bylaws be invoked to settle their accounts. The pertinent part of Section 5 read:

> "All indebtedness of a member shall be a lien upon the member-ship and upon the stock certificate evidencing the Resident Membership, standing in the name of such member, and such lien may be enforced by private sale of such stock certificate after ten (10) day's written notice to the delinquent member, stating the time and place of such private sale, such notice to be served either personally or by registered mail addressed to the most current address of the Resident Member as reflected by the records of the Club."

It was a rare occasion when someone was dropped from the Club roll. Most of those who couldn't pay their bills simply resigned. Some who couldn't and some who wouldn't first received a nice letter from Upshaw on special stationery which bore the names of the members of the Finance Committee, one or more of whom was generally a prominent banker. In friendly, respectful language, Upshaw would write that he was required to report the names of delinquents to the Finance Committee within 10 days. It was a task he abhorred, his letter implied, and a fate the delinquent did not deserve. All of it was a white lie. He was not required to report to the Finance Committee, and the stationery was of his own manufacture. The ploy almost always generated a spate of payments by return mail or even by special messenger.

If Upshaw's resourcefulness did not bring results, however, the next step was coldly calculated to draw the proverbial blood from the proverbial turnip. The delinquent's name was posted on the Club bulletin board. It was a devastating blow to the ego which prompted quick payment and, in may cases, wonderfully imaginative stories to explain away the tardiness.

In the beginning Upshaw had been hesitant to employ the bulletin board posting procedure. In his young manhood he had worked at a club where oldtimers had deliberately fallen behind in their bills because they reveled in the attention their names on the bulletin board produced! "Look at me!" they were saying. "I was a member of this club when most of you birds were in diapers!" But the method worked at the Petroleum Club of Houston. Few members were dropped from the membership roll for non-payment through invocation of Section 5, Article IX.

And no member, to this time, had been dropped from the roll for any other reason. This had come as a major surprise to Upshaw. He had heard that Texans, and oilmen in particular, were heavy drinkers, and he had expected a certain amount of trouble in this area. But no. After lunch the Club emptied like a schoolhouse at day's end, and seldom did anyone linger at the bar to drink away the blues. Indeed, the businessman in Upshaw was affronted when he learned that the Denver Petroleum Club, and other smaller clubs, had larger bar sales than his own.

At the February 9 meeting, after hearing from an outside expert on the remodeling plans, the House Committee was authorized to work out the remaining details and get the job done. At this meeting the three immediate past-presidents—Butler, Halbouty, and Bartle—were appointed to a nominating committee in anticipation of the upcoming annual membership meeting. The committee would select six candidates to fill Board vacancies, and the membership would elect three of the six.

It was brought to the Board's attention that the roof of the newly-constructed Ten Ten Garage, a seven-story parking facility at 1010 Travis, might be fine place to construct a new Petroleum Club. The location was desirable, and parking facilities were right at hand. And there was reason to believe the building owner was amenable to a proposal. It was decided to pass the information on to the Special Committee.

At the next Board meeting, on March 9, Sterling reported that Manhattan Construction Company was starting the remodeling work and should be finished within a month. Upshaw was instructed to dispose of furniture which would become surplus with the remodeling.

Upshaw, meanwhile, was approached by the Club waiters. They wanted better pay. They said waiters at the Houston Club, the Coronado Club, the Tejas Club, and the Ramada Club made more money than they did.

Upshaw knew this was true. He also knew that the Petroleum Club wage scale was based on what was generally regarded as a reasonably correct ratio between payroll and income. A pay increase would throw the ratio out of balance unless there was an accompanying increase in the volume of business. Nevertheless, he surveyed the other clubs to honor a promise to the waiters. Employees at the other clubs averaged about $20 more a week than Petroleum Club employees, he determined.

Upshaw met with Bagby, Selig, and Gueymard, chairman of the Finance Committee. They concluded that they could not and would not abandon the precepts of the founding fathers; they would retain their employees, who were excellent, and maintain a high standard of service. The would raise wages, not only for waiters but for other employees who deserved it.

At the April 13 meeting Upshaw presented his recommendations for wage adjustments to the Board. They were approved. Upshaw pointed out that use of private dining rooms by small parties, sometimes as few as three guests, was a problem. The luncheon tab usually totaled only slightly more than the wages of the waiter who served the parties. He was instructed to create a realistic price schedule for the private dining rooms for the Board to consider.

At the annual membership meeting of April 21, 1959, members voted to amend the bylaws to permit proxy voting. And almost immediately the ancient, enduring question of allowing ladies in the Main Dining Room at the noon hour was introduced for debate. James T. Mackey, an early member, pointed to declining Food Department revenue. Perhaps when the remodeling was completed, revenue could be increased by catering to the ladies, he said.

Be careful, said Hugh Q. Buck. Letting the ladies in for lunch might interfere with prompt seating and service of men with a limited time in which to eat. One of the Club's attractive features, he said, was that a member was seldom kept waiting to be seated and, once seated, usually was promptly served. Take away this advantage, Buck said, and the Club might very well lose many regular luncheon guests to other clubs and restaurants more centrally located.

Bill Bartle disagreed. With the remodeling completed, sufficient space would be added to the Grill to relieve the Dining Room of any over-crowding caused by the ladies.

And Carnes Weaver, an early member, said that perhaps allowing the ladies to use the Cocktail Lounge at lunch should be considered.

But no one generated enough enthusiasm to push for a vote and ladies at lunch returned to limbo.

Bagby, Hardin, and Selig stepped down as directors. Elected to succeed them were F. W. Bell, Joe T. Dickerson, and Raybourne Thompson. Bell, it will be recalled, was manager of Brown & Root's chemical department. Dickerson was president of Shell Pipe Line Company, and Thompson was a partner in the Vinson, Elkins law firm.

The next day, at the annual organization meeting, Howard Warren, the club founder, was elected president by secret ballot. Elected to serve with him were Walter Sterling, first vice-president; E. Clyde McGraw, second vice-president; Raybourne Thompson, secretary; H. F. Beardmore, treasurer; F. W. Bell, assistant secretary-treasurer.

According the the minutes, "Mr. Warren stated that, due to the untimely death of R. R. McLachlen, it would be necessary for the Board to appoint someone to fill his unexpired term which had one year to run. Mr. Warren said that the election for Directors showed W. J. Gillingham (of Schlumberger) in fourth place, and that he suggested consideration be given to appointing him to fill Mr. McLachlen's unexpired term. Michael V. Kelly made a motion that W. J. Gillingham be appointed. It was seconded by H. F. Beardmore and unanimously approved.

During the Hugh Q. Buck administration, limitation on resident memberships was lowered from 900 to 825. There were three chief reasons given for this action: The membership already included what might be called the pick of men who made up the oil and associated industries; the financial condition of the Club eliminated the need to supplement Club funds through the sale of additional memberships; the establishment of a waiting list would add further prestige to existing memberships.

Ignoring the first reason, the second and third soon lost whatever validity they may have had in the developing economic recession and a steady stream of resignations that would offset new memberships throughout the second half of the 1950s. There was never a waiting list because membership never reached 825. There were 810 members in 1955-56, 811 in 1956-57, 795 in 1957-58, 778 in 1958-59, 768 in 1959-60. Non-resident memberships also declined annually. And not even all the special events and heroic labor by Upshaw could prevent an annual loss in revenue in the food and beverage departments.

Club annual reports differ but little from the annual reports of large corporations. Both tend to accentuate the positive. This is not to imply that rank and file Club members were not made aware of Club problems, but bad news reached them only periodically, and it often was blurred by an abundance of glad tidings.

With money in its various bank accounts, the Club did not appear to be in imminent danger. But on the other hand, there was no reason to believe that the downward trend in revenues from membership sales and sales of food and beverages would soon be bottoming out.

Many informed members were concerned, but none was more concerned than Howard Warren. As a founding father, he had helped nurse the concept of the Club to reality. Over the years he had never stepped aside from a Club assignment. No one knew more about the Club than he. He was a warm, friendly man—the kind other men find it easy to talk with. "Pete," he would ask, "why the hell did you resign from the Club?"

"The location is bad, Howard," Pete might say. "Downtown traffic is bad and parking's worse. And my wife. For her to enjoy a downtown club, I've got to go somewhere else."

Later Warren would say to Ginther, "We've got to move, Wilbur. As soon as we can."

So far the search for a new site had been conducted at a leisurely pace in anticipation of the expiration of the Rice roof lease in 1966. Now, as president, Warren was in position to inject urgency into the hunt. "We've got to hurry," he told Ginther.

And that's what he told the Special Committee he appointed. The members were Don Connelly, Hugh Q. Buck, Butch Butler, Mike Halbouty, B. G. Martin, and Martin Sandlin. Warren chose Connelly to be chairman. Connelly, Warren told intimates, "will work around the clock if he has to." He had given Connelly a strong supporting cast. Now he told Connelly, "Find us a place, Don."

Meanwhile, he had retained Willard Gill as chairman of the Publications Committee. He had appointed W. J. Gillingham chairman of the House Committee, Herbert Seydler chairman of the Finance Committee, Frank Isenhart chairman of the Entertainment Committee, Samuel B. Symington chairman of the Food Committee, and Robert S. Moehlman chairman of the Library Committee. Isenhart, Symington, and Moehlman were new to committee chairmanships. Isenhart was purchasing agent for TGPL, Symington was a producer, and Moehlman was executive vice-president of Austral Oil Company.

With the remodeling work done, House Rule 3 was changed to read as follows:

> "Ladies are admitted to the Cocktail Lounge and Private Dining Rooms beginning at 11:30 a.m., and to the Main Dining Room at 1 p.m. Advance reservations are required for the use of Private Dining Rooms. The Gallery and West Wing of the Club are reserved for men until 5 p.m. after which the ladies will be admitted. On Sunday the ladies are admitted to all Club quarters during the hours that the Club is open."

The resolution was to become effective December 1, 1959. Only Walter Sterling voted against it.

Upshaw had told the Board that men began leaving the Main Dining Room at 1 p.m., and he thought that ladies could be served beginning at that hour with no inconvenience. Sterling apparently was not convinced. But F. W. Bell told the Board that many members were under the impression that alterations to the West Wing had been made primarily to provide additional space for the Men's Grill and thereby open some tables for the ladies in the Main Dining Room. So, for reasons of chivalry and economics—a lovely combination—the House Rule was changed.

Upshaw informed the Board that the AFL-CIO had begun a campaign to organize the club employees in Houston, and that several of the clubs had developed the idea of sharing the cost of legal counsel were legal counsel needed. Should the Petroleum Club share in such service and costs?

No, said the Board. The Club was loaded with legal talent. If labor difficulties arise, we'll handle them.

At the December 14, 1959 meeting the Board accepted with regret the resignation of Joe T. Dickerson. He was about to retire, Dickerson said, and he would be away from Houston much of the time. Warren moved that Ernest Barger Miller, Jr., a nominee for directorship at the past election, be appointed to serve Dickerson's unexpired term. Miller was a Tidewater Oil executive. Warren's motion was seconded by Michael Kelly and approved.

All during the year 1959–60 the Board handled the Club's routine affairs and took steps it considered necessary to improve the Club's financial position. The downward trend in revenues, however, was not arrested.

But almost incredibly, Don Connelly, in that brief time, found the membership a new home. As the founding fathers and charter members had yearned for the Rice roof, so did the present directors and members hunger for the pleasure dome that Connelly, by prodigious labor and keen negotiations, presented for their approval. At the annual membership meeting of April 19, 1960, it was announced that the brethren had voted 509 to 43 to depart the quarters they had once considered "the finest in the land."

A month after that decisive vote, the members had something else to think about—briefly. On June 10, 1960 the *Houston Chronicle* published a story under the heading, "Mad Barber Breaks Up Plush Club." The story began: "An unemployed barber made a tornadolike tour of the plush Petroleum Club atop the Rice Hotel Friday, leaving the elegantly decorated entrance lobby a shambles."

The *Press* story began: "A hypnotist charged earlier this week with unlawful practice of medicine today went on a rampage in the plush Petroleum Club, tossing valuable art objects through plate glass windows and terrorizing employees. . . ."

The intruder had been in the state hospital at Austin, John Sealy Hospital in Galveston, and Jefferson Davis Hospital in Houston for treatment as a chronic schizophrenic. After a trial on charges of felony malicious mischief for vandalizing the Club, he was found insane and sent to the state hospital in Rusk. A psychiatrist testified that the man had said he tore up the Petroleum Club "to get publicity" in connection with his earlier arrest for the unlawful practice of medicine as a hypnotist. Why he picked the Petroleum Club instead of some other prominent gathering place was not revealed.

The man showed up in the Club shortly after seven in the morning. No members were there and only a few employees. Before he was stopped by police he had broken three windows overlooking Texas Avenue, destroyed one of a matched pair of Venetian statues, demolished an ancient Chinese war god figure, turned over chairs, couches, tables, ash trays, and pulled down some draperies. Damage was estimated at $5,000.

The year 1959–60 was a year of decision for the Petroleum Club of Houston. It also was a year when two decisions were made that would dramatically change the course of oil history and have a profound impact on the world's economy. Both decisions were a result of the great crude surplus which, as we have seen, developed after the Suez Crisis of 1956.

The majors acted as a unit when dealing with the Middle East oil producers, but they approached the producers on a country-by-country basis. Nevertheless, on August 8, 1959, Standard of Jersey—without consulting either the countries or other corporations in the unit—cut the price it paid for Middle East crude by 10 cents per barrel.

The action infuriated the producing countries, who saw their budgets wrecked, and the other corporations in the unit, who felt betrayed by Jersey's unilateral decision. The corporations, however, had little choice but to swallow their anger and follow Jersey's lead.

But a month after the price cut, representatives of five oil producing countries—Saudi Arabia, Iran, Iraq, Venezuela and Kuwait—met in Baghdad, Iraq and formed the Organization of Petroleum Exporting Countries. Other countries in time would join the ranks. It was their intention to act as a unit, as the majors did, in oil dealings.

The infant organization was derided. Critics said that the political and religious rifts among the members precluded unity in oil matters. But OPEC would slowly grow in strength. And the day would come when it would rattle the world's economy like dice in a dealer's cup by withholding its crude from commerce.

Back in the States, the cries of the independents were louder than ever. In the first eight months of 1960, imports were 60,000 barrels per day above the average for the same period in 1959, and domestic production was down 77,000 barrels per day. The American Association of Oilwell Drilling Contractors—with members in the Petroleum Club of Houston—asked that imports be reduced and limited in 1961 to enable domestic producers to regain their share of the U.S. market lost since the end of the Suez Crisis . . . and that thereafter domestic producers be permitted to furnish the major portion of the increase in U.S. demand.

Apparently no one was listening.

But there were other things to talk about at the Club. Oilmen love "firsts" and "biggests," and there were many of them in the late 1950s. An early Club member, John Mecom, set a depth production record when he brought in a well that flowed from 21,465 feet in the Lake Washington, Louisiana field. A few months later he took one down to 22,570 feet. And then Phillips Petroleum drilled a wildcat to 25,340 feet in Pecos County, Texas.

Richfield Oil obtained the first real commercial production in Alaska with a wildcat drilled on the rugged Kenai Peninsula.

"Mr. Gus II," the largest drilling platform ever built (to that time), was commissioned in Beaumont. It could drill in water 150 feet deep and cost $6½ million.

The CATCO group claimed the first quadruple well completion in the Gulf, producing from four separate zones through four strings of tubing. So Sinclair promptly came up with a quintuple completion in the Gulf, and the first sextuple completion was claimed by Sun Ray Mid-Continent Oil Company in Jackson County, Texas.

Trunkline Gas Company used helicopters to string pipe across muddy terrain, and Texas Eastern constructed a 30-inch pipeline to the Mexican border to import Mexican gas on a large scale basis. Humble Oil and Refining Company drilled the world's largest gas well in Winkler County, Texas, the well's open flow potential being clocked at 1.66 billion cubic feet per day from 9,440 feet. Humble and Freeport Sulphur Company unveiled the largest steel island in the Gulf as a base for mining sulphur.

And in 1959 a new Humble Company was chartered when Standard of Jersey merged Humble, Esso Standard, and Carter Oil into the new organization. Very quickly the new Humble gathered in a number of smaller companies.

Before the merger with Esso Standard and Carter Oil and acquisition of the smaller companies, Humble had been the largest domestic producer of crude oil. Now it was America's largest domestic oil corporation—and it had grown too big for its britches. Its modest headquarters building at Main and Polk was not proper housing for such a giant.

Humble had plans for what would be the tallest skyscraper west of the Mississippi River, a 44-story tower that would rise 606 feet above street level and more than 100 feet higher than any other building on the Houston skyline. It would consume an entire city block bounded by Travis and Milam Streets and Bell and Leland Avenues. At its southwest corner, also occupying a city block, would be a six-story garage with 1,200 parking slots.

This gleaming white tower would be seven blocks south of the city's central business district and 10 blocks south of the Rice Hotel. At hand would be a belt of freeways and other traffic arteries that would make the building site easily accessible from any section of the city. An air-conditioned sub-surface passage would connect the building and the parking garage.

But the skyscraper was still a dream, and the building on Main Street was bulging with personnel. One major problem occupying the minds of Humble's directors was the executive dining room. It was considered one of the finest in the city, but now it was at overflow point with the staff attempting to feed as many as 300 persons at lunch.

And the renowned chef, having reached retirement age, had retired.

CHAPTER NINE

HUMBLE HAD LONG been a Texas institution, occupying a special niche in the hearts of Texans. It was the thousands of service stations with their "Happy Motoring" logos that brightened street corners and country roadsides. It was Kern Tips and his fellow announcers broadcasting Southwest Conference football games over an Humble football network—as certain a sign of autumn as turning leaves. It was a great refining complex at Baytown, Texas where country boys from Polk County and elsewhere could get a full day's pay for a full day's work and build a future. "By God," they would tell their friends back home, "Old John Humble works you hard but he treats you fair, and if you can't work for him, you can't work for nobody!" This was the grandest kind of compliment, word-of-mouth publicity more valuable than any kind of purchased advertising. Humble employees, whatever their craft or profession, were accorded respect because of the general acceptance that Humble was an honest and stable company.

In the oil patch, Humble was regarded as a fierce competitor.

The company was founded in 1917 by men who got oil in their veins during the great Spindletop boom of 1901. They came from diverse backgrounds and brought a variety of skills and properties into the organization. Each was outstanding in one or more phases of the oil business, and the organization was destined to prosper with this rare blend of talent and ambition.

The first board of directors was composed of Ross Sterling (president), W. S. Farish, H. C. Wiess, R. L. Blaffer, W. W. Fondren, Frank P. Sterling, C. B. Goddard, Attorney L. A. Carlton, and Jesse Jones. Jones was seated because of his reputation for business acumen, his ability to borrow money in New York for the fledgling outfit, and a $20,000 investment. He sold his stock and resigned from the board in less than a year. He was replaced by E. E. Townes, Carlton's law partner. The attorneys had represented most of the founders separately, and had drawn up the company's incorporation papers.

Sterling got his start as the owner of a chain of feed stores he set up in Sour Lake, Batson's Prairie, Dayton, Humble, and other early fields. His brother, Frank Sterling, was his partner. The stores provided fodder for the horses and mules that pulled the oilfield wagons of the era. The feed business flourished, and the brothers acquired oil interests in Humble, Goose Creek, and Oklahoma. The Texas properties were wedded under the name Humble Oil Company with headquarters in Houston.

Farish had been a young lawyer in Mississippi when a client sent him on an errand to Beaumont at the height of the Spindletop boom. He did his job, but oil fever had grabbed him. In Beaumont he became active in drilling and, perhaps more importantly, studying ways and means of making the budding industry a vital and continuing part of American life.

Farish met Blaffer in a Beaumont rooming house. Blaffer had come from New Orleans to buy oil for the Southern Pacific Railroad. He stayed to join Farish in contract drilling and trading leases at Spindletop and elsewhere. They later moved to Houston to concentrate on the Humble field.

Fondren and Goddard came to Spindletop as roughnecks; Fondren by way of the Corsicana field, Goddard by way of the Lima, Ohio field. Both became drilling contractors, and both became wealthy. Fondren, particularly, acquired large production in several fields. He was widely regarded as a "master driller."

Wiess, a native of Beaumont, entered the oil business as an investor. But he was a born organizer, and his Paraffine Oil Company not only discovered Batson field but pioneered at Humble field and established commercial production in the North Dayton field. He also formed Reliance Oil Company and found success in Oklahoma.

These individuals operated independently. It was Farish who proposed, in 1916, that they pool their production to guarantee a steady supply of crude to potential markets. The others liked the idea, and early in 1917 they consolidated their properties. In June they received a charter as the Humble Oil and Refining Company.

Another founder was J. Cooke Wilson of Beaumont. Wilson did not become active in the fledgling company, but he contributed his oil production for stock. He remained in the business as an independent, starting the Wilson-Broach Oil Company, later to be run by his sons, J. Cooke, Jr. and Waldo. (When J. Cooke, Jr. was accepted as a member of the Petroleum Club, he listed his occupation as "roustabout.")

Humble had grown into a big integrated company by the time Dad Joiner discovered the East Texas field in October 1930 with his Daisy Bradford #3. Like the other big companies, Humble had written off the area as barren after its geologists and geophysicists had found no indication of structures suitable for oil entrapment. But when a second well, the Lou Della Crim #1, roared in near Kilgore on December 28, Humble went into action.

The great well had been drilled by a wildcatter's wildcatter, Ed Bateman, on a 1,494-acre lease. Bateman was so broke when the well came in that he complained that he "couldn't make a down payment on a free lunch." Nevertheless, he told Humble that he wanted $1,500,000 cash and $600,000 in oil produced for his seven-eighths interest in the well and the lease it sat on. He never wavered from those figures during an all-day session with the Humble board of directors in Houston.

The Depression was in full swing and Bateman was asking for a lot of money. But Humble's geologists had studied the well and had become convinced that Bateman's price was a pittance in comparison with the wealth beneath his acreage. On January 9, 1931, Humble closed the deal with Bateman.

Seventeen days later—as if to confirm the geologists' assessment—the Lathrop #1 was brought in near Longview, and Humble's landmen sallied forth to battle with hordes of wildcatters for East Texas acreage . . . and they got their share.

Humble would be successful in other areas in other years, but East Texas assured the company of great reserves. It was commonly said in the oil patch that "East Texas made Humble."

W. S. Farish was the Humble president when the Bateman deal was consummated. Ross Sterling had left the company in 1925. He was elected governor of the state in 1930, and took his seat on January 10, 1931—the day after Humble met Bateman's price.

Blaffer and Wiess would later serve as Humble presidents.

Standard of Jersey had acquired 50 percent of Humble's stock 'way back in 1919, and later increased its shareholding to 75 percent. Humble regularly supplied the Jersey board with directors, and Farish would become a distinguished Jersey president.

Humble was no more loved by the independents than any other major company during the turbulent East Texas boom years. As for Governor Sterling, he served only one wild and woolly term, being defeated in his reelection bid.

Humble prospered during the forties and fifties and, with the 1959 merger, became the largest domestic oil company. With, as we have seen, a large housing and feeding problem.

Now Humble had virtually ignored the Petroleum Club of Houston, a slight that particularly had stung the founding fathers, the charter members, and the early members whose extra push had brought the Club into being. Humble had purchased no company memberships for its executives, or anyone else. Only two Humble employees were card-carrying members of the Petroleum Club, and they had joined as individuals.

Humble had its own fine executive dining room to offer as an excuse for not supporting the Club, but it never officially offered it. And if it had, Club members would not have accepted it. The Club, in their eyes, was much more than a mess hall for executives. Much more. It was the only club for oilmen in the oil capital of the world. And it was the finest in the land.

Humble's directors were in a position where they would soon have to decide *which* ranking officers could eat in the company executive dining room. It was suggested that the company buy memberships at various clubs in the city and dole them out to various executives. It would be a stop-gap measure until executive offices and a dining room, planned for the two top floors of the new building, were ready for use.

The suggestion found some acceptance because Humble's directors had voiced concern in the past that company officials spent too much time with each other. Getting some of them out in the world would broaden their experience.

"Perhaps we should talk with the Petroleum Club," someone said. "They might be interested in the two top floors of the new building. We could install our executive offices somewhere else in the building and make dining arrangements with the Club for a good number of our officers."

It was pointed out that the Club's lease on the Rice roof still had a number of years to run, and that a remodeling job had just been completed.

"I have reason to believe they're ready to move," the first director said. "Having the Petroleum Club on top of our building would lend us prestige and be attractive to potential tenants."

It was something to think about.

It was something for Don Connelly to think about also. He was highly regarded in Humble executive circles, and he heard about the Humble board discussion indirectly while the subject was still being considered by Humble officials.

Like Mohammed, Connelly was not the kind to sit around and wait for the mountain to come to him.

Don Connelly was born in the shadow of the oldest, and one of the richest, oilfields in America—the Bradford field of northwest Pennsylvania. It was discovered in 1871, 12 years after Colonel Drake's well at Titusville set off the world's first oil boom. Long after other fabled Pennsylvania fields were history, Bradford continued to yield crude of such exceptional quality that crudes from other fields in the country were measured against it. A hundred years after its discovery, Bradford had produced more than 650,000,000 barrels, and was still producing, at a greatly reduced rate, in the 1980s.

Connelly was born at Bradford in 1898. His father was an independent producer in the field. Connelly got his first taste of the oil patch while still a boy, roughnecking on his father's cable tool drilling rigs. After attending Catholic University in Washington,

D.C. and Columbia University in New York, and serving as a lieutenant of infantry in World War I, he headed west, lured by the great oil strikes in the Mid-Continent.

His first job was with National Supply Company in Bigheart, Oklahoma. He was an attractive, cocky young man with a thirst for knowledge and a talent for making friends. His desire to learn—and an itching foot—kept him on the move. He left National Supply to work for Tidewater Oil Company in Tulsa, and then for Gulf as scout and landman. He was an independent operator at Cisco, Texas for two years before going to work as a landman for the old Indian Territory Illuminating Oil Company of Bartlesville, Oklahoma, which later became part of Cities Service Oil Company. Cities Service sent him to Houston as manager of the Land and Geological Division. In 1944 he left Cities Service to head up the Oil Division of Warren Petroleum Corporation. He held that position in March 1949 when he joined the Petroleum Club of Houston.

It will be recalled that Bob Smith chose Connelly to serve on the Club's first Advisory Planning Committee during and after construction on the Rice roof. The membership had honored him with election to the board of directors in 1954 and he had served under three presidents, Buck, Bartle, and Halbouty. And he had handled a number of important Club assignments.

Now, in the 1959–60 Club year, he was a major executive of Warren Petroleum. He was known for his good works, particularly in connection with the Catholic Church. In 1955, for example, he was general chairman of a successful drive to build four new Catholic high schools in Houston. And in 1956 he had been invested as Knight Commander into the Equestrian Order of the Holy Sepulchre of Jerusalem. His wife Martha had been honored as a Lady Commander in the Order. Connelly would be blessed with other Church honors in the future, and locally he would be recognized for his civic and religious enterprise.

As he undertook the assignment to find the Petroleum Club a new home, he was a vigorous 61 with more than a trace of the cocky, attractive young man who had ventured out from the country's oldest oilfield to find out what was going on in the burgeoning Mid-Continent. He had become a resourceful, respected executive in the world of oil.

Three years earlier, in 1956, Hugh Q. Buck had measured a vacant lot at Travis Street and West Dallas Avenue, and had offered a suggestion that the club could buy "sky rights" in advance of construction at the site. Alvin Moody had spent a great deal of time discussing such a possibility with potential builders at the location. Moody later had brought up the idea of the Club constructing its own building, much as David Bintliff had advocated such a program before the move to the Rice roof. Moody even drew up a cost sheet for "Petroleum Club Tower Building, Inc." His plan called for a 10-story structure—eight floors to be rented out, two top floors for club use.

Later Halbouty, Buck, and Upshaw held serious discussions with the operators of the Ten Ten Garage. The location had its attractions, and talks were still going on when Connelly assumed chairmanship of the Special Committee. He immediately joined in the talks.

And he asked Upshaw to provide him with a breakdown of the estimated space needed should the Club move to new quarters. Upshaw did. "On the basis of my own calculations, it would appear that we would need approximately 40,000 square feet," he wrote. "We now occupy 22,714 square feet, and our rent is $60,000 a year. We have a fifteen-year lease which expires in November 1966. The lease includes an option to sub-let to an acceptable tenant. It does not include a renewal option."

Attached was a breakdown of the estimated space needed, with a cautioning note: "It should be understood that this is an estimate to be used as guide to help in determining how much space we would need; it is not intended as a plan."

What followed was a dream that had been edited by a supreme realist . . . four tightly-written pages that would have thrilled an architect and set an interior designer to humming. Upshaw had taken that estimated 40,000 square feet of space and had lovingly squeezed every utile inch from it. It was as if an admiral had ordered him to draw up plans to invest an imaginary island whose dimensions were given but not the topography, climate, or the number and physical characteristics of the natives. "In the main dining room," he wrote Connelly, "seating capacity and area required is based on 16 sq. ft. per person for regular service. This allowance includes space for service tables, tray racks etc.

12 sq. ft. per person are allowed for banquet service. In the Men's Grill, 15 sq. ft. are allowed for regular service and 12 sq. ft. for banquet. . . ."

Room by imaginary room he explored, and reported his findings to Connelly. "The members have not used the library to a great extent in the past, and it has been said many times that we do not need one. However, for a Club of our stature, I feel that space should be alloted, somewhat as a retreat, for members who desire a quiet place to relax and read. 625 sq. ft. would provide such a room. . . ."

On and on he wrote until he had exhausted the potential with 12,000 square feet left over. "This breakdown," he concluded, "does not take into account the possibility of a card room or health club, nor does it include lobbies or hallways. There are approximately 12,000 sq. ft. unassigned which may be sufficient to serve these needs. Necessary passageways for members and for service can be determined only after the general layout of the Club is known. . . ."

Armed with such a formidable weapon, Connelly sallied forth to talk with Sheraton Lincoln officials who were planning a hotel for downtown Houston. He continued to talk with people from the Ten Ten Garage. He went to see James A. Elkins, Sr., whose First City National Bank was building new quarters. Elkins said no. He wanted a smaller club in his building—the Ramada Club. Petroleum Club members would overtax the elevator system they were so numerous. "But Don," he added, "if you find a place and you need money to make the move, come and see me."

It is likely that Connelly was shadow-boxing with these maneuvers for by now he had talked with an old friend, Harry Ferguson, an Humble vice-president. He had learned that the Petroleum Club was *wanted* on the two top stories of the skyscraper-to-be. With this knowledge in hand, he confidently called on Morgan Davis and Carl Reistle, Jr., Humble's president and executive vice-president, respectively. He was greated cordially and assured that Humble was prepared to make the Club an attractive offer. His Humble contact, he was told, would be Ray H. Horton, an Humble vice-president and president of Humble's producing division.

From that point on Connelly began operating as stealthily as a secret agent. Only Upshaw had his complete confidence. Early in August Connelly and Horton arranged for Upshaw to fly to Los Angeles for a conference with Welton Becket and Associates, Humble's architects.

The architects showed Upshaw blueprints of the Humble building. The 43rd and 44th floors provided more than 40,000 square feet of net space, as he had hoped for. Becket was well known as a designer of famed clubs, and the architects showed Upshaw examples of their work. What Becket had done for others was fine, Upshaw had to concede, but nothing he saw was suitable for the Petroleum Club of Houston. Upshaw felt that he knew what his membership would expect if it elected to move from the Rice roof. He made a number of suggestions in that direction, and he left for Houston with the architects' promise that a set of blueprints would be mailed to him.

At this point Connelly casually let Warren and his fellow committeemen know that had talked with Humble officials, as well as representatives of other sites, but gave few details. However, in a memo dated September 1, 1959, he was a bit more revealing. The memo said:

> "Since so many of you travel and are not available and I have happened to be in the city the last couple of weeks, I have had two meetings with the Sheraton groups as follows:
>
> "1. Last week I met with Lloyd Bentsen, Jr., of the Lincoln Liberty Life which is sponsoring the Sheraton hotel project in Houston; a Mr. Mills of the Sheraton group, from Boston, and the architect. I told them about the space contemplated if and when we decide to move from the present quarters, and outlined some general ideas that we have all discussed heretofore.
>
> "2. Today I met with Messrs. Hoghn, of Boston, and Davis, of Dallas—both with Sheraton—to further discuss the Petroleum Club possibilities for the Sheraton Lincoln Hotel, if and when it is erected.
>
> "The point is that they have not decided as yet as to the date for breaking ground, nor do they have the actual plans for the type of building or the number of square feet per floor. They are thinking, however, of having the top floors contain 13,400 sq. ft. each. This is just a short memo to bring you down to date and advise

you of the contacts. I have had meetings with Tennessee Gas for the Ten Ten Building and, as you know, with the Humble. Both are to submit architectural ideas which have not as yet reached me. Mr. Upshaw has also made a trip to California to talk with the Humble architects.

"I shall be glad to call a meeting anytime anyone suggests, and also be glad to receive any ideas from each of you. Of course these contacts with all concerned are confidential."

The next day Upshaw received the prints from the Humble architects—they were not at all to his liking—and shortly thereafter Connelly set up a meeting with Warren and the Special Committee members for September 10.

After reporting on his meetings with Sheraton officials, Connelly finally got down to business. According to the meeting minutes,

"Mr. Connelly mentioned that at the suggestion of Mr. Horton of the Humble Oil and Refining Company and Mr. Welton Becket, the architect for the new Humble Building, Mr. Upshaw had gone to Los Angeles to discuss in some detail with the architects what our needs for new quarters would be; and that Mr. Upshaw had received some drawings which indicated that the 43rd and 44th floors would meet our requirements.

"Mr. Connelly said he had recently talked further with Humble officials and that they gave him further assurance that they would like to have us as tenants. He said that he felt he had gone as far in his discussions with Humble as he could go until the committee members expressed themselves as either favorable or unfavorable to the Humble location. If, he said, the committee was of the opinion that the new Humble Building would make a good location for our new quarters, he could then approach Humble for an offer. He felt that before making a report to our Board of Directors he would like to be able to give them sufficient information regarding approximate costs, etc. for them to be able to decide whether or not this particular project should be pursued further.

"After considerable discussion, which covered many aspects pertaining to the subject, it was the unanimous opinion of those present that Mr. Connelly should have further discussions with the Humble authorities, seeking whatever information he deemed necessary for his report to the Board of Directors.

"The Committee then looked over the preliminary drawings of the architects and also a revision by Mr. Upshaw. It was their opinion that the drawings showed the possibility for one of the most outstanding clubs in the country. A number of changes were suggested which included enlarging the proposed Main Dining Room where banquets up to 500 could be served, and perhaps enlarge the Men's Grill to seat 200. . . ."

On September 29, 1959, Upshaw wrote a letter to Charles R. Sikes, Jr., the Becket architect in charge of the Humble Building-to-be.

"With the basic information you included, I was able to work out what appears to me to be a functional use of the space according to our needs. I am enclosing a set of amateur drawings which present a graphic picture which incorporates a few changes in our original thinking. . . ."

The drawings may have been rough, but they were decidedly professional. What Upshaw had done amounted to a virtual redesigning of the allotted area.

"The position of the Main Dining Room was changed because the present shape has a number of advantages over a long narrow room as shown on your drawing. . . .

"The Men's Grill, located on the southeast part of the floor, seems to offer several desirable advantages. We have placed the Cocktail Lounge on the northwest side which places it in convenient proximity to the Main Dining Room and at the same time removes it from the Main Loggia. . . .

"All private dining rooms have been placed on the 44th floor. We think the esthetic effect of the double ceilings in the corners on the northwest and the southwest part of the building are quite desirable, but we feel that the loss of floor space at the opposite end on the 44th floor offsets the value of the effect in that area. Therefore, we wonder. . . .

"Returning to the Main Dining Room—it is unfortunate that the columns do not allow for a dance floor with an unobstructed view of an orchestra platform, or in the case of banquets, the speaker's table. . . .

"I am somewhat concerned about the size of the Kitchen space as compared with the dining area. . . ."

And Upshaw concluded his letter thusly:

> "I am sure you realize that at the moment we are trying to determine whether or not the space being considered can be developed into Club facilities that will meet our needs. We will be most appreciative of the comments of you and your associates."

Sikes knew that Humble wanted the Club on top of the tower. Upshaw was saying, in effect, "we're still just thinking about it." But Upshaw well knew that Connelly and the rest of the committee members would be satisfied with no other location.

Connelly finally reported to the Board of Directors on November 9. Howard Warren, of course, was privy to what Connelly had been doing, but if any other director had the faintest inkling it was not apparent in their reaction to his disclosures.

Connelly went through the rigmarole about the Sheraton Hotel and the Ten Ten Garage and his talks with James Elkins, Sr. of First City National Bank . . . and Humble.

According the minutes,

> "He had found Humble very interested. He was now anxious to have an expression from the Board Members as to whether or not they regarded the Humble site with sufficient favor for him to pursue the matter to a point where he could ask Humble for cost figures insofar as a lease arrangement was concerned; his approach being that it was our desire to have the landlord turn over the Club facilities in accordance with our own desires, and lease it to us on the basis of their investment in the property. It was understood that furnishings and equipment would be furnished by the Club.

> "The Board of Directors was very impressed by the possibilities of our Club being situated on the top two floors of the Humble Building, and asked Mr. Connelly to discuss with the Humble people the terms and cost to the Club of leasing the property. The Board also expressed its appreciation for the very excellent work done by Mr. Connelly and his Committee. . . ."

The circle of enlightenment was expanding. Connelly went back to work. So did Upshaw. If Connelly was the catalyst of the contemplated move, Upshaw supplied much of the elbow grease. For Connelly he met with the Humble officials who were directly responsible for the building and agreed on basic terms of who would supply what in the club's construction. For Warren and the directors he made a study so he could "present to you information of reasonable reliability which can be used as a guide in determining the Club's ability to carry the additional burden of higher rent on larger quarters such as those proposed atop the new Humble Oil building. . . ." (Modest almost to the point of reticence, Upshaw in later years would attempt to minimize his role to an interviewer, but the Club records clearly reveal his prodigious efforts. At one point Director Walter Sterling wondered aloud if the Club was not placing too much new responsibility on a man who also was expected to do a first-class job of managing the current Club.)

Ground was broken for the Humble building on February 16, 1960. Two days later Connelly prepared a memorandum for his files "because someone soon will succeed me as chairman of the committee and will take over all negotiations if it is decided to continue further with the Humble."

Referring to his talks with Ray Horton of Humble, Connelly noted,

> "These conversations have taken place over several months, and by their nature they have had to be more or less confidential. The only person who has been completely informed—other than myself—is L. A. Upshaw, Manager of our Club . . .

> "The Board offered and passed a resolution authorizing the continuance of negotiations along the lines which I outlined (to it), and which I will not mention in this memorandum because they are still of such a confidential nature there is no occasion to have general information passed to members of the Club until something concrete and definite has been agreed upon. . . .

> "It has always been my idea that the present members of the Club should not pay for and install required portions of the building for club facilities that would result in a large outlay of cash currently, and would react to benefit of members down through the years. With that thought in mind, I have been approaching Humble on the basis that Humble be responsible for permanent improve-

ments such as refrigeration, rest rooms, painting, decorating, floors, etc. . . .

"To install these items will require an outlay—according to architects and building management—of between $600,000 and $700,000.

"Humble wishes to reserve about 1,800 sq. ft. of space for three small dining rooms for noon use exclusively out of the 41,850 sq. ft. contemplated for Club's use, and have us pay for about 40,000 sq. ft.

"Ray Horton and I have discussed price per sq. ft. and he has not made a commitment to me because he will have to appear before his Board for authorization; however, we have both been bandying back and forth a price of between $3.50 and $4.50 per sq. ft. Horton will not go below $4 for recommendation to his Board, and he hopes next week to recommend to that body leasing to us the space for $4 per sq. ft. . . .

"Again, I reiterate that is not a firm price from Humble as it must be passed on by the Board. Horton has suggested $165,000 per year, which slightly exceeds $4 per sq. ft. It was also mentioned that if we assume the cost of the items mentioned, covering some $600,000, the space would be rented to us at a rate of $130,000— which is a bit more than $3 per sq. foot. The term of the lease discussed is 20 years, with renewal rights. . . .

"I have also talked about a guarantee of 50 memberships from Humble. Humble presently, I understand, has two memberships."

Later in the day Connelly met with Warren, Upshaw, and members of his committee—Buck, Butler, Halbouty, Martin, and Sandlin. He passed out his memorandum. Pointing to the $4 per sq. ft. Humble had suggested as a fair price for rental of the 40,000 square feet available, he said that comparable space in Houston rented for $5 to $6 per square foot. And, at that price, Humble would furnish a great deal in construction not normally furnished by the landlord—between $600,000 and $700,000.

That, he said, was about what it would cost the Club to finish, furnish, and equip the quarters. And Upshaw said the club would have about $500,000 in its Sinking Fund by the time the money was needed two years hence.

Halbouty said: "We have to move to establish ourselves in the area in which the city is growing. On top of that skyscraper, we can have a club of such beauty it will attract new members, it will stimulate old members, and allow us to handle larger parties and banquets. This added business will offset any added costs." He moved that Connelly get a firm offer from Ray Horton, and the motion was carried unanimously.

Marlin Sandlin had been through this before. He stressed that the Club should "employ our own architect and interior decorator at the same time in order that the whole program of design and decorating can be coordinated from the very start. . . ." It was sage advice, as events would demonstrate.

All present were handed a copy of a report from the Club's accountants, Harris, Kerr, Forster, and Company. It was a projection of operating costs, expenses, and operating income in the proposed new quarters. It differed but little from the study Upshaw had done for Howard Warren. The Club could make the move and operate with no trouble if the move prompted increased membership and increased business, the report indicated.

Let's do it, said the wildcatters.

If Connelly was a man who liked to move right along with a project, so was Ray Horton of Humble. On March 1, 1960 he sent Connelly a proposal that he wanted signed by the president of the Petroleum Club of Houston. In an accompanying letter, however, he brought up a delicate subject he and Connelly had discussed at length. Horton didn't think the subject should be a part of the proposal he wanted Howard Warren to sign. He wrote Connelly:

> "We have discussed from time to time the matter of Humble employee memberships in the Club. After a thorough discussion of this question we are of the opinion it is in the best interest of the Club and of the Humble Company not to make as part of our agreement a commitment to purchase a definite number of memberships should the Club move to the new Humble Building for several reasons, the two most important of which are:

"1. We do not want to leave the impression that the Humble Company is attempting to dominate the Club or force it to allocate to us a definite number of memberships.

"2. Purchase of the memberships would require a policy consideration for the Humble Company with regard to the assignment of memberships."

It should be remembered that Connelly joined the Petroleum Club of Houston in March 1949. He was aware of the trials of the founding fathers and later the charter members in trying to get the major companies to aid the Club through company memberships. He could recall vividly that the majors—with the exception of Gulf—had turned a deaf ear to all proposals. It was his thinking that if Humble would buy, say 50 company memberships, it would encourage other major companies to do likewise. So he had made Humble's purchase of 50 company memberships part of his bargaining.

Now Horton wanted him—and the Club—to take such an arrangement on faith.

"All of the members of our present Company Board of Directors as well as our Division Board of Management have indicated their desire to become members of the Petroleum Club if it moves to the new Humble Building; and, as I told you, I am confident that approximately fifty of our top executives and perhaps more will desire memberships.

"It would seem to me, however, that it would be advisable from both parties' standpoint for Humble employees who desire memberships to apply as individuals and their memberships be processed in the same manner as other applicants. We would of course hope that those who apply would get favorable consideration. If this procedure were used, criticism should be held to a minimum since the Club would have complete freedom to accept or reject individual applications on individual merit. . . ."

Connelly was in total agreement. He told Howard Warren how he felt about the unwritten pledge in a brief letter mailed March 3, after Warren had the Humble proposal in hand:

"At this point, I wish you would emphasize to the Board that there is not a member of either the Humble Delaware Board or the Humble Division Board who is not a first-class citizen in every respect, and I can vouch for them because I know them, even those who have recently moved here from New York and Tulsa. Our dealings have all been on a high plane; Humble wants no advantages and I for one am proud that the situation has developed to the proposal which is now in your hands. . . .

"I am for it and I hope that you can prevail upon the Board to accept the proposal and authorize your signature to the letter submitted by Humble . . .

"I think the salvation of the Club is in this move, and I also think when it is a "fait accompli" all members will be tremendously pleased. . . ."

Warren presented copies of the proposal to the directors at the March 14 Board meeting. The proposal listed the items to be furnished by Humble and the items to be furnished by the Club. Rent would be $165,000 a year on a 20-year lease with two renewal options of 10 years each.

In the proposal was a matter Horton and Connelly had discussed at length. Horton wanted to retain 1,800 square feet on the 44th floor for Humble's exclusive use as a dining area for executives and selected personnel. The Club would cater the dining area at its customary rates. Connelly wanted Horton to guarantee that Humble would buy a fixed number of meals each week. In the proposal Horton estimated the number at 750–800 weekly. The idea later was abandoned for a variety of reasons, the chief one being Humble's belated recognition that the plan negated the desire to expose ranking officials to the outside world.

Warren said he though the Board should determine whether it had reached a point where it was ready to vote on rejecting the proposal or recommending it to the membership.

Walter Sterling wasn't ready to make that decision. He moved that Warren call a joint meeting of the Board, the Special Committee, the Finance Committee, and the Advisory Committee to review the proposal. The motion was seconded and approved.

The joint meeting was held on March 21. Sixty-two committee members were in attendance. From the meeting grew a letter that Warren mailed out to all members on March 30. Much of the letter's contents have been covered, but what is said and left unsaid merits its complete inclusion in this story.

"As most of you know, for the past several years we have had an active committee appointed for the purpose of planning for the future welfare of the Club. It became apparent some three years ago that because of the rapid growth of the city in the direction South of our present location, we would eventually find our Club no longer conveniently situated. This view has been borne out by the fact that each year we have had an increasing number of resignations and less frequent use of the Club by our members.

"The Special Committee became convinced that as we approach the lease expiration date on our present quarters, it was imperative that a new location be found. It, therefore, set out to explore possible sites with certain objectives in mind, principally a site that would remain a choice and convenient location for years to come; facilities of sufficient flexibility to allow greater participation in activities and events of special interest to the Petroleum Industry, good parking conditions and, of considerable importance, provisions for our wives.

"During the past year, Mr. D. L. Connelly, Chairman, and the members of the Special Committee have had a number of conferences with possible landlords of the several sites considered; the most attractive, financially and otherwise, has been the offer made by the Humble Oil and Refining Company to provide space for Club quarters on the 43rd and 44th floors of their new office building which is to be constructed on Travis between Bell and Leland streets.

"The adjacent 1,200 car garage would solve parking problems that have always existed at our present location. The feeder streets and freeways nearing completion and others on the drawing board, make this location easily accessible for the approximately 40% of our membership whose business addresses are beyond the limits of downtown Houston. The building owners intend to lease about 40% of space to non-company tenants, and undoubtedly many of our members will have offices in the building. With the new Sheraton-Lincoln Hotel now under construction nearby, and the vast new Cullen Center soon to be constructed, there is little likelihood of our finding a location or building that has so much to offer.

"It is with this background that the Special Committee unanimously recommended to your Board of Directors that the Petroleum Club accept Humble's offer of the space mentioned for new Club quarters which will become available in about two and a half years. The Board, having had numerous discussions on the subject, desired to present it to the Advisory and Finance Committees for consideration. On Monday, March 21, a joint meeting of the Advisory Committee, Finance Committee, Special Committee, and Board of Directors was held in the Club quarters, and a very thorough discussion of the subject took place. Sixty-two committee members attended the meeting.

"The Club was indeed fortunate in having a number of members attending who were well acquainted with downtown properties and real estate values. They advised us that the offer made by the prospective landlord was so attractive that it was doubtful another equal opportunity would come our way.

"Aside from the fact that we would be provided with quarters equal to any in the country, it is not contemplated that the cost will add any additional burden on the members. This unusual situation was made possible by the founding fathers of the Club when they established a sinking fund requiring a portion of dues to be set aside for our future needs.

"The following resolution was, on motion of Thomas H. Wheat, seconded by W. L. K. Trotter, passed without expressed dissent:

'Resolved that the proposed Club site on Top of the new Humble Building be and it is hereby recommended to the full membership for approval, on such terms as the Board of Directors of the Petroleum Club may approve.'

"Enclosed is a proxy and a return envelope. It is important that every member check his vote and sign the proxy. Mail it or bring it to the Club at your earliest opportunity. It must be in the hands of the voting judges prior to the Annual Meeting on April 19, 1960, at 4:00 p.m.

"We are convinced that your affirmative vote will be in the best interest of the Club."

It was a strong letter. Things are not going well, it said. Vote yes and things will improve tremendously. Believe us.

In the meantime the Board had named the three most recent past presidents—Bagby, Halbouty, and Butler—to select eight nominees of which four would be elected to fill vacancies on the Board at the Annual Membership Meeting.

Connelly was vacationing in Florida when he received a phone call from Butler. "Look, Don," Butler said, "you've been on the Board before, but we're going to nominate you again. And you're going to win the election. And when the new club opens in sixty-two, Martha (Connelly's wife) is going to cut the ribbon because you're going to be president."

Politics and power plays were not unknown in Petroleum Club activities.

At the Annual Membership Meeting of April 19, 1960, Warren announced that the vote for moving to the Humble Building was 508 for and 43 against.

He asked if there were any members who had not cast their vote by proxy and, if so, did they now wish to vote.

Up spoke A. L. Selig, long-time member and former director. He hadn't voted because Warren's letter hadn't said anything. How much was the move going to cost? What was Humble asking that wasn't in the letter? What about the lease and its terms?

He couldn't vote intelligently, Selig said, because he didn't really know what he was supposed to vote on. "I want to know about the so forths and so ons," Selig said. Implied was that the others who had voted were willing to buy a pig in a poke. (Every club, every government, needs its Seligs and Logan Bagbys.)

So Connelly took the stand and reviewed the work of the Special Committee and outlined Humble's terms and cost of the lease. Herbert Seydler, chairman of the Finance Committee spoke on the money angles. Increased membership and increased business would take care of things. The Club's cash reserves would approach $600,000 by the time the members were ready to move. "But everybody is going to have to help in building up membership," Seydler said.

Said Selig, "All of this information would have served a good purpose had it been included along with the proxy." Whereupon he cast his vote in favor of the move.

Whereupon, Warren said the overwhelming favorable vote was authorization for the Board of Directors to conclude an arrangement with Humble, and that it would do so at the appropriate time.

Then, according to the meeting minutes, Warren announced that

> ". . . the dues for the present and subsequent quarters would be exempt from federal tax if they were used to construct and furnish the new Club quarters. He stated that 100% of the dues (after Beverage Pool deposit) could be used for such purpose, because the Club has sufficient funds on hand to cover any deficit in current operations, and that, by taking advantage of the tax exemption, payment of the usual amounts by members would result in 20% additional funds being available to the Club for new quarters. This matter was discussed by the members."

The discussion resulted in a resolution that annual dues and taxes thereon henceforth would be placed in a separate bank account and be spent only for construction and furnishing the new club.

With that out of the way, Warren called for a report on the results of the election of four new Board members to replace Gillingham, McGraw, Sterling, and himself.

The report was read—and lo! Don Connelly's name led all the rest.

The other new directors were A. G. Gueymard, Harry Hurt, and A. W. Waddill. Gueymard and Hurt had been active committeemen. Waddill was a founding father.

At the annual organizational meeting the next day Beardmore was elected president. Kelly and Dean were elected vice-presidents, Gueymard was elected secretary, Seydler was elected treasurer, and Waddill was elected assistant secretary-treasurer.

Beardmore told his fellow board members that he would make a courtesy call on the current landlord to inform him of the Club's decision to move.

It seemed to be the proper thing to do.

CHAPTER TEN

HERBERT BEARDMORE was an honors graduate of the University of Oklahoma with a geology-petroleum engineering degree in his pocket, but it took more than that to get him in the oil business and keep him there. It took pride, and a rare kind of courage.

It was 1925. Beardmore was 21, a tall, slender young man standing on a drilling rig floor in the Ranger field in Eastland County, Texas. He had reported for his first oilfield job, as a roughneck, and the driller was pointing straight up to the "monkey board," high in the derrick.

Beardmore had drawn on an inner strength to be able to clamber up on the rig floor for he suffered from acute acrophobia. Now the driller was ordering him aloft to "work the derrick" much as early seamen climbed the rigging to furl and unfurl sails.

Great fear of heights can soften brain and bone and resolution. It cannot be laughed away or prayed away. But from some reservoir Beardmore summoned up the nerve to climb the derrick, step by tormented step, until he reached the "monkey board" and with trembling fingers tied his safety rope around his waist.

And he worked there, doing his job, fighting off the demon that threatened to yank him off his perch. He went back the next day, and the next, and the next, never erasing his terrible fear but proving bit by bit that he could control it. Legends have grown from lesser struggles.

Within a year Beardmore was practicing his profession with the Indian Territory Illuminating Oil Company in Oklahoma City. From 1935 to 1943 he was a district engineer for Amerada Petroleum Corporation in Tulsa. Like many other engineers who worked for Amerada, Beardmore fell under the influence of Charles V. Millican, Amerada's great engineer who pioneered the study of bottom-hole pressures. Beardmore became recognized throughout the Mid-Continent for his work in devising methods to handle high-pressure gas in the lifting of crude.

He was chief engineer for Barnsdall Oil Company for four years before joining Warren Petroleum Company as production manager. He was made a vice-president in 1955, shortly before Warren became a Gulf subsidiary. In 1958 he was named manager of Gulf's Houston Production District, and in 1960, the year he was elected president of the Petroleum Club, he became vice-president—U.S., in charge of all of Gulf's U.S. exploration and production activities. He was the first ranking officer of a major company to head the Club. He had joined in 1949.

Beardmore's first act as president was to sign the letter of intent that Humble's Ray Horton had sent to Howard Warren by way of Don Connelly. His second was to telephone Fred Heyne, Sr. of Houston Endowment, the club's current landlord, to inform him of the anticipated move from the Rice roof. There is no record of that conversation, but it is likely that Heyne already possessed the information and, inferring from future events, it also is likely that he was politely noncommittal in his reply.

To consummate the deal with Humble, Beardmore retained Connelly as chairman of the Special Committee with Hugh Q. Buck and B. G. Martin as his aides. And he appointed a New Quarters Committee with the authority to employ an architect and interior designer and to supervise construction generally until the new club was completed. It was a big job, a most responsible one, and to chair the committee Beardmore selected A. G. Gueymard. Wilbur Ginther was named vice-chairman. Members were George T. Barrow, F. W. Bell, Gaston C. Jones, R. B. Mitchell, Robert S. Moehlman, and Marlin Sandlin.

A. W. Waddill was named chairman of the House Committee; Amos A. Roberts, Entertainment Committee; Wallace C. Thompson, Finance Committee; E. Porter Johnson, Food Committee; W. J. Mechura, Publications Committee. Roberts was a Baroid executive, Thompson was president of General Crude Oil Company, Johnson was a Sun Oil Company executive, and Mechura dealt in oil properties.

At the annual membership meeting it had been decided to place all dues and the tax on them in a special Capital Improvement Fund. Now Beardmore pointed out to the directors that the Board had the power to change membership and transfer fees from time to time by simple resolution. He moved that future fees plus taxes also go into the Capital Improvement Fund (CIF), and it was approved by unanimous vote.

It was obvious that assignment of dues to the CIF was going to reduce the Club's operating income substantially. Michael Kelly moved that the loss be made up by drawing on the Club's reserve fund when necessary. This too was approved unanimously as was a motion by Harry Hurt that the CIF be set up in the Bank of the Southwest.

Wallace Thompson and his Finance Committee recommended that the CIF funds be invested from month to month in six-month deposit certificates paying 3 percent interest, and a motion to do so was approved by the board.

More money would be needed for building the new home, however, and that could only come with an expanding membership. There would be a lot of discussion about getting new members in the following months, but little would be accomplished.

Meanwhile, Amos Roberts and his Entertainment Committee members had enlisted their wives' aid in planning future programs. The ladies, happy in the new freedom granted them with the change in the House Rules, promptly began planning a Ladies' Bridge Luncheon. They conferred with Upshaw.

The minimum luncheon menu price was $1.85, Upshaw told them.

Too much, said the ladies. A lighter lunch was more desirable, and $1.50 per person should be sufficient.

Ummm, said Upshaw.

Now, about cocktails. Paying for downtown garage parking and lunch would make it a rather expensive party. Guests certainly would expect cocktails. Could a Vodka or Rum Punch, for example, be included in the $1.50 lunch, the ladies asked.

Upshaw stalled until he could put the question before the Board. Furnishing cocktails as part of such a modestly-priced lunch would be tantamount to subsidizing the bridge party, he was aware, and contrary to Club policy. He explained the problem to the directors.

According to the Board minutes,

> "In a general discussion by the Board members it was their feeling that the price structure of food items must be maintained at a level that will keep the department on a sustaining basis, and that the price structure on services in the Beverage Department should be kept at a level which would allow this Department to operate at a profit since it is the only sales department we have that can contribute substantially toward the operating costs of the Club. To deviate from our past policy would set a precedent that would require different price structures for different groups, which would be an unwise move. Therefore, Mr. Upshaw was informed that we must conform to our past policies. . . ."

Thus it was left to Upshaw to tell the ladies that if they wanted cocktails, they would have to pay for them.

And the Food Committee happily reported that a newly-installed roast beef buffet in the Men's Grill was being well patronized. It was popular because it was accessible. At that time only the employees were aware that the buffet was popular for another reason also. While a diner was telling the chef how rare or well-done he wanted his slice of beef, he also would indicate how thin or thick he wanted it . . . and considerably more thick slices were served up than thin ones. So many more, in fact, that the beef buffet was contributing less than believed to operating revenue. Upshaw quietly alleviated this minor problem by showing the chefs how to miscalculate to the thinward on the more obvious too-thick orders.

After that slight correction the Food Committee could definitely report that "the increased and continued popularity of the roast beef table has enabled us to obtain better food and serve more members at less cost. . . ."

The Food Committee was a "working committee," an appellation that distinguished active committees from those that weren't. E. Porter Johnson, the committee chairman, quite regularly offered the Board excellent suggestions to improve food and service. At least one, however, was ahead of its time. The Committee recommended to the Board that a Food Department Manager be hired with the sole responsibility of supervising the production and service of food.

Don't do it now, Upshaw told the Board. The club simply couldn't afford the $8,000 to $9,000 a year a qualified man would demand, and a qualified man couldn't carry his own weight. The volume of business at present and in the immediate future couldn't justify such an employee. "When we approach our time of moving we'll want such a man so he can get acquainted with our way of operating, and so he can have ample time to develop his own organization." And he added suavely, "With the good Food Committee we have, and a staff that is anxious to make day-to-day progress, we can get excellent results without any substantial increase in payroll costs. . . ."

A Board so sensitive to Club economics that it could deny the ladies free cocktails at a bridge party could not disregard Upshaw's mild lecture. It decided to delay action on the Food Committee's recommendation.

Beardmore had asked that the committees be "working committees," and they were. He knew that it was important for the Club officers and committees to strive a little harder before the move to the Humble building. He was wont to drop in at committee meetings, inspiring the members with his interest in what they were doing.

Amos Roberts and his Entertainment Committee did such a good job that members regularly commented to the Board that the "such and such" party "was the best I ever attended."

Roberts' report to Beardmore and the Board near the end of the year demonstrates the scope of the committee's interest, and also gives an idea of the work past Entertainment Committees had done since Marlin Sandlin had appointed Michel Halbouty to chair the first one. Roberts wrote:

> "During the Spring and Summer our program of entertainment consisted of a series of Buffet and Dinner Dances.
>
> "With the return of members from summer vacations, our Fall program began with a Neiman-Marcus style show. Throughout

the football season chartered buses were run to and from Rice Stadium. Special pre-game Cocktail and Buffet Suppers were held at the Club.

"For the first time, the Club held a party honoring our new members. Mr. and Mrs. Ed Farrell were in charge of this function and are to be commended for its splendid success.

"An Election Night Party provided an occasion for something different at the Club, and it was a complete sellout. The Club was decorated to represent campaign headquarters for both parties, and prizes were awarded to those guessing the winning candidate's electoral votes. It was January before winners could be determined. The success of this party was largely due to John Heinzerling and Robert E. Souther, ably assisted by all members of the committee.

"Mrs. Stanley Ward and Mrs. Ivan Burrell did an excellent job with the Ladies' Bridge Luncheon. The turnout for this party filled the Cocktail Lounge, and we believe the interest warrants several throughout the year, particularly after moving to our new and larger facilities.

"The Committee planned an early Christmas party which featured the music of the Season. The Club was filled with members and their guests, who enjoyed joining the Seventh Day Adventist Carol Singers in making the evening a pleasant one.

"Another special was the Mexican Supper Night. We selected late January for this event, proving that members were well rested from the Holiday Season and ready for a good party. Hats and favors, along with the decorations and beautiful buffet, made this a real Fiesta.

"The Spring Fashion Show by Sakowitz Bros. introduced us to what we can expect for our ladies for Spring and Summer. An added bonus for the evening was the modeling of Men's Fashions by our Club members. Wilbur L. Ginther, Michael V. Kelly, Charles H. Dresser, Jr., and George Pierce cooperated in this portion of the program.

"This report precedes our New Orleans Party scheduled for April 15, which promises to be a gala event. The entertainment is being imported from New Orleans, and will be interesting to our members who are familiar with the French Quarter.

"Activities such as the Golf and Gin Rummy tournaments have become regular events. James I. Riddle and Stanley Ward were in charge of this part of the program.

"Good attendance for the special events indicates that the membership has enjoyed the entertainment program throughout the year,
and your Committee is pleased to have had the pleasure of serving in this capacity.

"Your chairman wishes to express his appreciation to all Committee Members, and especially to our wives who did most of the
work."

The wives, indeed, contributed much to the program's success,
and future Entertainment Committee chairmen would bless Roberts and follow his example of enlisting the ladies.

Connelly, Buck, and B. G. Martin were busy roughing out a
lease agreement with Humble's Ray Horton, quibbling amicably
with the Humble man over an occasional snag. Horton couldn't
guarantee the use of a specified number of parking places for Club
members and couldn't agree to reserving a specified number at
any certain time, but he gave his solemn assurance that members
would be taken care of properly. Connelly accepted it.

Horton agreed that Humble would pay for the power and
chilled water for operation of the air conditioning system, but hot
water and steam for the kitchen would probably require a special
boiler and he thought the Club should pay for that.

And they debated about who would pay the maintenance and
operating costs of the passenger and service elevators from the
43rd to the 44th floor.

Connelly's fellow directors gently urged him to get on with it.

Shortly thereafter Connelly reported to the Board again, and
this time he called on Buck to brief the directors. Buck was in
charge of the legal problems involved. Like Walter Sterling, Buck
was regarded as an elder stateman. He was a charter member, a
past-president, and he had been steering the Club through legal
shoals almost from its inception. Buck delivered a brief lecture to
the Board.

There were a few items to be ironed out, Buck said, but the
Club was dealing with a cooperative landlord and the lease, when
ready, would be good one. But the directors must realize, Buck
said, that in working out an agreement in which landlord and ten-

ant must share in construction costs, it was impossible to anticipate and include every question that may develop. "Certain matters will require discussion and negotiation as construction proceeds." So let's do the best we can on getting the lease ready to
sign, Buck said, recognizing that there will be problems from time
to time until the work is completed.

The lease was ready to sign in mid-August 1960, and on August
22, in a simple ceremony, Beardmore and Horton signed it.

Gueymard and his New Quarters Committee had not been
idle. An architect had been hired, and so had an interior designer.
The committee had so thoroughly immersed itself in the planning
for the new quarters that the Board, with support from the Advisory Committee, had authorized Gueymard et al a budget of
$1,536,000 "to follow through to completion the new club atop
the Humble building."

Problems major and minor would plague the committee before
it could really get in gear. For example, Beardmore told the Board
that the United Gas Company was "particularly concerned with
the decision to equip our kitchen with electricity." The gas company's concern, he said, was that if an all-electric kitchen was installed in the Petroleum Club of Houston, "considerable adverse
publicity would be created on a national basis, particularly in the
trade journals."

After a general discussion it was agreed that the Board certainly
didn't want United Gas to get all that adverse publicity. What to
do?

Upshaw played Solomon. He was aware that Houston Lighting
and Power Company had employees in the Club, and that the
company used great quantities of gas and oil to fire its generator
boilers. But United Gas employees also were members, and the
Club basically had been established for those in the oil and gas
industries.

According to the Board minutes,

> "Mr. Upshaw stated that he was certain that we could install gas
> cooking ranges, broilers and roasting ovens, which constituted
> the major equipment, without materially affecting the operation.
> However, he did feel that the electric deep-fat fryer and electric
> bake ovens had a decided advantage over the gas equipment, and

he recommended that we not change our original plans in relation to these particular items."

The Board sighed, and agreed.

Two nagging worries continued to vex Beardmore and the Board throughout his term. The first was their inability to increase membership. Beardmore, like many others, had hoped that the mere announcement of the club's intent to move to the Humble Building would bring new members flocking to the fold. It didn't. James Clark, the public relations man, used his skills to publicize the Club's attractions and entertainment events. This brought old members back to the Club—and helped the budget—but it attracted no newcomers. An advertising firm, Craig and Associates, was engaged to upgrade the Club's mailings, and did it. But resignations still nullified new memberships.

Beardmore met with the Advisory Board and suggested that a membership campaign of some kind be launched. The Advisory Board shook its collective head. Such a campaign would be undignified. Beardmore, reporting to the Board, said wryly that the Advisory Board's belief that increased membership would come in good time was heartening, and he wished he could share its optimism, but he still believed it was better to do something instead of nothing. A step or two was taken, but they were tentative at best, and nothing of consequence resulted.

At the end of Beardmore's term the Club was still shy a dozen members in meeting the 825-member limit. Beardmore, however, chose to take the positive viewpoint. To him the Club was *only* a dozen short of the membership limit. Therefore, he recommended that the limit be raised to 1,100. The Board resolution that was mailed out to the membership reflected Beardmore's attitude. It said: "The active resident membership in this Club shall be limited to 1,100, exclusive of non-resident conversions and the clergy, and non-resident membership shall be limited to 750." The members voted for it.

The second nagging worry was Fred Heyne, Sr. and Houston Endowment. Heyne had been told on the telephone by Beardmore that the club was going to move in late 1962. That was all. As the months passed, there was a lot of talk at Board meetings about dealing with Heyne, and Beardmore eventually appointed a committee to handle the matter. What with one thing and another, the committee didn't get in gear right away.

Meanwhile, a consensus had developed among the Club's movers and shakers that the Club should be forgiven the rental on the Rice roof from the planned move-out in 1962 until expiration of the lease in 1966. At $60,000 per year the "forgiveness" would amount to about $240,000. It was justified, some felt, because the Club had made such a big investment in the premises in construction and decorating costs. And it was felt that it would be a nice gesture on Heyne's part, a civic gesture, no less, to let the Club off the hook.

Finally, Beardmore went to see Heyne. Before Beardmore could state a position, Heyne told him that he liked having a nice club on the Rice roof. "I hope you have luck in subleasing to a nice club," he said. In other words, he wanted to be paid in full and expected it to be done. Beardmore wisely retreated. Heyne would not be approached again until more than two years had passed.

The terms of the Beardmore, Dean, Kelly, and Seydler expired on April 19, 1961. Elected to succeed them were Charles W. Alcorn, Amos Roberts, Wallace Thompson, and Vernon Frost. Alcorn was vice-president of Falcon Seaboard Drilling Company. For Frost, the founding father, his election marked the second time he had been chosen as a director.

The Board elected Ernest Miller, Jr. president. F. W. Bell and Raybourne Thompson were elected vice-presidents; Amos Roberts was elected secretary, Frost was elected treasurer, and Wallace Thompson was named assistant secretary-treasurer.

Miller selected Stanley Ward of Tennessee Gas Transmission Company to chair the Entertainment Committee; Wallace Thompson, Finance; T. F. Palmer of Sinclair Oil and Gas, Food; Robert Moehlman, House. Gueymard, of course, continued to chair the New Quarters Committee, and Don Connelly continued as chairman of the Special Committee.

Miller was aware that Beardmore had wanted to select a New Members Committee but had refrained from so doing in deference to the Advisory Committee's belief that a major membership drive would be undignified. Miller didn't ask the Advisory Committee for guidance. He appointed Harry Hurt chairman of the New Members Committee, told him to select his aides, and urged him to begin work as soon as possible. He left to Hurt's good judgment how the work should proceed.

Miller's grandfather had been a pioneer Pennsylvania oilman. His father also had worked in the Pennsylvania fields before moving to Oklahoma in 1903 where he became a successful drilling contractor. Miller was graduated from the University of Oklahoma in 1930—the first year of the Depression—and for four years he knocked about the oil patch at various jobs, some better than others, until he landed a roustabout job with Tidewater in the East Texas field. He would stay with Tidewater his entire working life, climbing the ladder rung by rung. At the time of his Petroleum Club presidency he was vice-president and general manager of Tidewater's Southern Division. He had joined the club in 1957.

Miller led the Club through a busy year. Harry Hurt's New Members Committee planted posters extolling the virtues of the new club in the lobby of the old one so that visitors could catch the idea of what the promised land would be like. Robert Moehlman, the Austral Oil executive and member of the New Quarters Committee, directed preparation of a brochure on the new quarters for Hurt's use. Hurt began a campaign to get every member to act as a salesman. At year's end resident membership had grown from 820 to 972, leaving only 128 vacancies.

And Charles Alcorn of the Special Committee found a prospective tentant for the old quarters. The Ames Brothers, entertainers, wanted to operate a night club on the Rice roof. Negotiations continued throughout the year. At first the Club took a firm stand, wanting very much to get $60,000 a year for the quarters, which would allow the Club to break even on its payments to Houston Endowment. The Ames Brothers were tough bargainers, however, and month by month the Club dropped its asking price. At year's end, the two parties were still haggling.

Horton of Humble agreed to buy 50 company memberships in staggered amounts, beginning in October 1961, with all applications paid for by September 30, 1962. As noted earlier, Humble formally abandoned its plan for an executive dining room on the

44th floor, and turned over to the Club the 1,800 square feet of space it had reserved at no additional rent. The Club agreed to reserve a private dining room seating up to 20 persons for Humble's use daily from Monday through Friday, and Humble agreed to pay for a set number of meals each day whether or not the room was used.

There was only one minor dispute between landlord and tenant. Humble wanted the Club to pay about $150,000 for what Humble considered electrical wiring in excess of that outlined in the contract between the parties. It was finally settled with the Club paying half of the bill.

Miller kept in close touch with the New Quarters Committee and the two men it had hired to create the new club—Architect George Pierce and Designer William Parker McFadden. Pierce and McFadden had been engaged only after exhaustive research on the part of the committee. As John Staub and Edward Perrault had worked together to create the "finest club in the land" on top of the Rice Hotel, so Pierce and McFadden joined their skills to build a club "as new as tomorrow, as timeless as time."

CHAPTER ELEVEN

ARCHITECT GEORGE F. PIERCE was notified of his selection to build the new club on July 13, 1960 in a letter written by the New Quarters Committee Chairman, A. G. Gueymard. The letter would serve as an inspiration to the architect. It was a revelation of Gueymard's personality.

"We feel that an opportunity exists for the creation of an outstanding facility for the use and enjoyment of our members and their families, and we believe that your firm can capably effect this creation. We do feel, however, that we should acquaint you with our primary desires at this time so that you may govern yourself accordingly. Our feelings about the new club can best be expressed by the following criteria:

"1. The club facilities must be completely functional, and the design must result in sound operational economics. As you well know, this is a private club and is not subsidized by any form of government or civic organization and, therefore, it must survive upon its own merit of efficiency and desirability.

"2. The design and appointments must create a feeling of strength. Petroleum and its derivatives is primarily responsible for the progress and strength of this nation. The men who created our domestic petroleum industry were of strong character and it must be expected that those who will cause it to be continued must exemplify this same strength. It is believed that their surroundings should support and be in keeping with their efforts.

"3. We live in a world of rapid changes and those who change with it are considered progressive. Oil moves the world and is most responsible for the progressive decade in which we live. Being progressive, therefore, exemplifies the contemporary and we feel that this new club and its facilities should be of that order. Even though oil expresses the contemporary, it is also one of the oldest substances usable by man. In keeping with the age of oil, we would think that the contemporary aspect should be of the sub-dued type with a strong suggestion of age and maturity.

"4. Although the club is to be basically for the male sex, we must not completely disregard our opposite sex. They must be considered and provided for in those parts of the club appropriate to them. We feel that the use of the club by the ladies will contribute substantially to its economic success.

"5. In the matter of cost and expenditures of funds, we believe that the limitation should only be that cost which results in a facility that exemplifies good taste, function, ability and an expressiveness of enduring charm.

"The above outline of criteria, we believe, expresses the deep concern and feeling that the members of this committee have in this matter. We do not pretend, at this time, to outline the many details and instructions which will come later, but, instead, only desire to express our deep convictions.

"At this time an interior decorator has not been chosen. We reserve the right to choose an interior decorator to our liking provided, of course, that you feel capable of working with the particular decorator selected by us. In all probability the Committee will desire that we hire the decorator directly and that he be instructed to work with you. We anticipate that the decorator will be contracted for within the next month.

"We should like to have you feel, in assuming the task outlined, that you will always have the complete support of this Committee. We want to be kept informed of your progress and will make ourselves available at all times for conferences and discussions. We look forward to a relationship which will be gratifying to all concerned. . . ."

(The man who would be selected for an interior designer also was inspired by the letter, but he was an ebullient soul and he blithely remarked when he read it, "I get his drift. The club represents the best of the past and the present. Why try to outdo everybody else with a new pipe-stem chair? Isn't that what he's saying?")

The banker who wrote that letter was first an oilman. He came from a Louisiana family of farmers and merchants, and as a youngster rode with his father and an uncle to watch the rigs running at nearby oilfields. By the time he entered Louisiana State University in 1931 oil, in his words, "was just coming into its own," so he "signed up for petroleum engineering."

In his junior year, Gueymard's father died. It appeared he would have to leave school to help run the farm operations. But the family decided to shut down the farms, and Gueymard remained in school, working for the Baton Rouge Electric Company to help pay the bills. He stayed on at that company after he got his degree in 1935. He was surveyor, and he didn't like it. He made his way to Beaumont, Texas and got a job as a student engineer with Stanolind Oil. That meant he was sent to Barber's Hill as a roustabout, assistant mule-skinner, and the derrick man on a rod and tubing crew.

After that apprenticeship he was sent to High Island as a junior engineer and then to Houston as a "proration" engineer. But Gueymard wanted to work where the wells were being drilled, so Stanolind sent him back to Barber's Hill as a field engineer.

Then in 1938 Tidewater offered him a job of overseeing the drilling of some deep wells in South Louisiana. He was still with Tidewater when the Japanese bombed Pearl Harbor. A month later Gueymard was in the U.S. Army.

The night before D-Day he landed in Normandy aboard a glider; he was a captain of an anti-tank company of the famed 101st Airborne Division. He fought in France, Holland, Belgium, and Germany. He was at Bastogne during the Battle of the Bulge. He came home in December of 1945 as a major with four European campaign stars; two bronze star medals; two Presidential unit citations; Division Certificate of Merit; Dutch Orange Lanyard; Belgian Fourragere 1940; Belgian Croix de Guerre with Palm.

Back home, Tidewater made him an assistant division superintendent headquartered in New Orleans. Like many combat veterans, Gueymard was restless. He couldn't seem to adjust to what he was doing. So he quit his job, without prospect of another, and headed for Houston. He dropped by the office of an old friend, George Nye, an official of then First City Bank but formerly Tidewater's chief engineer. Gueymard told Nye he was looking for a job.

"Come on and go to work here," Nye said.

It was that simple. Gueymard began as an evaluation engineer. As time passed he made the transition to banking and would become a senior vice-president heading the petroleum and mineral division of First City National Bank.

Nye had joined the Petroleum Club in 1948. Gueymard joined in April 1949. His sponsors were Nye and Harris Underwood, Jr., the club's first president. Gueymard had been an active member over the years, always ready for an undertaking. He was, in Upshaw's words, "a powerhouse." He was a friendly man, but tough when it came to seeing that the Club's interests were served as the new home was created. And he was fortunate in that the men who served on his committee—Ginther, Barrow, Bell, Jones, Mitchell, Moehlman, and Sandlin—were as dedicated to excellence as he. "Each one was a joy to work with," he would tell an interviewer many years later. "We met at least once a week, sometimes several times, and they all attended. All were interested, and all contributed. They made it the nicest extra-curricular job I ever had."

Of the half dozen architects the committee interviewed, Pierce made the strongest impression. His presentation, the committee reported to the Board, showed a lot of thought, was more detailed than the others and, of greater importance to the committee members, offered a plan "to move people about, both members and employees."

Pierce was born and reared in Dallas. He studied physics for three years at Southern Methodist University before he admitted to himself that he really wanted to be an architect. The creative artist in him had finally overcome the precise scientist, though his strict regard for detail would be well recognized in his chosen profession. He transferred to Rice Institute in Houston and there received the degrees in architecture he wanted. Later he studied and did graduate work at Fontainebleau's School of Fine Arts.

So he knew what he was about when he went for himself in Houston shortly after World War II. He had been captain of the golf team at Rice, so he joined Houston Country Club. Affluent members—the kind who want buildings built—were happy to have a "scratch" golfer in their foursomes. And he delivered in the architecture his artistry and love of detail. He did work for Rice Institute, for SMU, and a number of oil companies. ("He has a fine reputation," Gueymard told the Board.) Welton Beckett, architect on the Humble building, had engaged Pierce as a consulting architect on the project because of Pierce's knowledge of the local climate and local mores. So Pierce already "knew" the building.

The first thing Pierce told Gueymard after he was hired was that something had to be done—quickly—that would cost about $70,000. The building hadn't been designed to have a club on top of it, he told Gueymard. Ceilings would be too low. Columns would impede sight and movement. The 43rd and 44th floors would have to be totally redesigned to lift the ceilings and move the columns out of the way. In fact, Pierce said, the building would have to be redesigned from 10 floors below the club so the columns could be moved and beams cantilevered. Construction crews were working apace and a decision must be made immediately. "It's a rush situation," he told Gueymard.

Gueymard explained the problem and the need for speed to the Board. He and Pierce were authorized to treat with Humble and Welton Beckett. The Club, said the Board, would pay the cost of the changes. Humble and Beckett agreed to alter the building plan, and Pierce was then free to design the spacious quarters he had conceived.

The designer, Bill McFadden, was chosen primarily because of his work on the recently-completed Chaparral Club in the Southland Life Building in Dallas. "Outstanding," the committee reported to the Board. He also had designed the interior of the City Club of Baton Rouge, a club with which Gueymard was familiar. Pierce also was impressed by McFadden's work and reputation. "He's a great conceptual design man," he said, "and he knows the sources for materials and he knows the artisans."

Pierce was in his early forties, full of vigor and ambition, and he recognized that producing a club satisfactory to the membership would propel him into a brighter future. McFadden, on the other hand, frankly told Gueymard's committee that he expected the Club to be his crowning achievement. So both men were highly motivated, and both were fired with a competitive spirit. As Gueymard would say later, "Both were good, but both wanted to be first. They were battling for first place all the time. We were aware of the competition, but it never caused a problem, and the Club profited from it. And I believe they came through it all with an increased regard for each other."

McFadden was talkative, high spirited, and he seemed to be on intimate terms with prominent people around the world. Some club members—and Upshaw—thought at first he was a name-dropper. But when McFadden, Pierce, and Upshaw journeyed around the country to look at famous clubs or to locate artisans and art objects, it was as if someone had placed a notice in the newspapers to announce that McFadden was coming to town. "Everywhere we went, someone of prominence was waiting to throw a party for old Bill," Upshaw reported. "He was full of love, and people responded by loving him." And Gueymard told the Board, "He knows exactly what he wants, exactly where to get it, and darn near exactly what it will cost."

McFadden's work style was demonstrated when he told the Committee that the Club Pierce was designing called for "big things rather than a lot of little things." On a wall in what would be the main dining room, the Petroleum Room, he wanted to place a huge tapestry depicting some phase of the oil business. He was delighted when several members of the committee simultaneously suggested that the tapestry depict a geological cross-section of Texas from the Gulf Coast to the Permian Basin.

Gueymard appointed Robert Mitchell and Robert Moehlman to work with McFadden. Moehlman, a hard-rock geologist with a Ph.D. from Harvard University, went to his office from the meeting and sketched the idea on an 8″ × 11½″ sheet of paper, coloring it in earth tones as McFadden had specified. Almost, that is. McFadden grabbed his head when he saw that Moehlman had colored the area near the Gulf Coast a beautiful blue. "Because of the Gulf waters," Moehlman said. McFadden wouldn't have it. "Earth tones all the way," he insisted. Moehlman made the change. "You're right," he told McFadden.

Mitchell, also a geologist, looked over the sketch. He made a suggestion. A. L. Wirt, a respected geologist not on the committee, studied the sketch. He made a suggestion or two. Finally, McFadden got the sketch, had an artist named Carlos Calderon check out the colors, then flew to the French Riviera where a Houston artist of international renown, David Adickes, had a studio at Antibes.

Adickes bought some pastel wax crayolas, each as big as a cigar, tucked the sketch in his pocket and set out for Madrid. At the shop of Real Fabrica de Tapices he explained what he wanted. He was supplied with a huge sheet of paper which he sliced into eight panels. He spread the paper on a workroom floor.

On his hands and knees, with the sketch beside him, he painstakingly drew the sketch to scale!

The job took him two weeks. Eight weavers took a panel each and hung them in their work areas. Placing their looms in front of the panels, the weavers wove their magic, using 287 specially-dyed woolen yarns to translate the sketch into a magnificent tapestry.

Each weaver in one day could weave about as much as a man could cover with a hand!

It took about a year to complete the tapestry, the largest ever woven in the ancient establishment. Adickes would drive from Antibes to Madrid once a month to inspect the weavers' work. When they were through, Adickes gave the weavers his blessing and arranged to ship the tapestry to New York.

The U.S. Customs office in New York notified Upshaw that the Club would have to pay a $2,000 duty on the tapestry. "It is not a work of art," the Customs office said, "and hence not duty free." The tapestry was so big it had to be a commercial property, the Customs office said.

Attorney George Barrow, a member of the New Quarters Committee, took charge. Using enough legalese to let the Customs officials know they were dealing with a professional, Barrow wrote a four-page affidavit describing the tapestry from conception to completion. His writing style verged on blank verse and was as full of color as a rainbow. If the tapestry wasn't a work of art, the affidavit implied, then neither was the Mona Lisa.

All right, said the Customs office. Come and get your work of art. Duty free.

Throughout the Club, Pierce and McFadden had specified that the padded leather on the walls and chairs be especially created to reflect the iridescent quality of petroleum. In a state teeming with cattle, McFadden insisted that the desired leather must come from the hides of Scottish beasts. There were three reasons for his insistence, he explained. First, big hides were needed, and the cattle of Scotland were larger than their American cousins. Second, thick hides were needed, and because of Scotland's cooler climate, nature provided the Scottish creatures with a thicker, hairier covering. Third, and most importantly, Texas and other U.S. cattle, by inadvertence or an understandable desire, were constantly pricking their hides on the barbed-wire fences that enclosed them. Scottish ranchers and farmers, on the other hand, penned in their cattle with stone and wooden fences. "No blemishes on the hides of the cattle of Scotland," McFadden said.

Hides from Scottish cattle, by all means, said the committee, and McFadden went to Scotland. He bought the hides he liked, and had them shipped to a tannery in New Jersey where they were treated and dyed. Some then were shipped to Houston to be mounted on the Club walls. The remainder was shipped to Copenhagen. There they were used in the upholstering of chairs designed by Hans Wegner and built by Johannes Hansen. McFadden modified Wegner's design so that the chairs were suitable for "big-butted Texas oilmen."

Gueymard's committee had got an early start, and Pierce and McFadden were able to add the finishing touches to the new home once the construction crews sheathed their tools and took off for other jobs. The membership had hoped to open the Club officially on December 15, 1962, the 11th anniversary of the opening of the old Club, but a brief construction strike and other customary snags that crop up in the erection of a giant building caused some delay.

Meanwhile, the Club had made a very bad deal with the Ames Brothers, sub-leasing the old quarters to the entertainers for what amounted to little more than a month-by-month rental. And the venture quickly failed. Fortunately, Upshaw had made a deal whereby Petroleum Club members were allowed to sign tabs at the night club and the charges were placed against their Petroleum Club accounts. When the Ames Brothers could not pay their rent, Upshaw did not pay the Petroleum Club members' night club tabs, so the Petroleum Club suffered only a minimal rental loss. The Ames Brothers had agreed to buy some of the Petroleum Club furniture, and Upshaw took it back.

Upshaw and Attorney Barrow went to the old Club to let the Ames Brothers know that time had run out. It was night. They conferred with the manager and agreed to let the club stay open until midnight before they turned the key and locked up the place. They were ushered into a private room and served drinks. "Uppy," Barrow said, "We can't have too many of these or we'll forget what we're here for."

Some of the furnishings in the old club were moved to the 44th floor of the new one. Some time later the Board authorized Upshaw to sell some of the furniture to John Mecom, an early Club member. Mecom also wanted to buy the beautiful Emperor Josef porcelain stove. Upshaw knew a bargain hunter when he saw one. Mecom could buy the stove if he also bought the seven-foot Kwan Yin statue. "It's a package deal," Upshaw said.

Mecom bought. (The stove now sits in the Warwick Hotel on Main Street, the statue stands in the Warwick-Post Oak, both controlled by Mecom.)

So the Club had the old home back on its hands.

The Humble Building contractor told the Club he would turn over the quarters in mid-January 1963, and McFadden said he would need another month to perfect them. So the opening date was set for February 16.

A pamphlet was prepared to be given to the membership shortly before the opening. It provided a guided tour of the club.

"The lighting has been intricately designed to provide glare reducing interior illumination during the day and beauty at night. Material textures and colors have been designed in some cases to reflect light and, in others, to absorb it. Also important has been the use of materials in such a way to provide acoustical balance.

"The various facets of the petroleum industry are reflected throughout the Club. In the principal areas, earth colors are used. The leather finishes . . . reflect the iridescent quality of petroleum and other minerals. The carpet design on the 43rd floor was inspired by striations of marble. Fossil forms have inspired various designs. Various metals have been incorporated in furnishings, wall plaques, and sculpture. The coloration in the glass globes in the chandeliers is related to the opalescence of petroleum, and the design of the pewter finished chandeliers was inspired by the early forms of oil burning lamps. (George Barrow asked a visiting New York lawyer what the chandeliers reminded him of. The lawyer glanced upward. 'Wagon wheels,' he said. Barrow chuckled. 'So much for art.')

"Many outstanding artists and artisans have contributed to the Club decor. The shaded inner drapery is shot with vertical threads of copper and pewter tones and was handwoven by Maria Kipp of California. (Kipp was another of McFadden's favorite artists, and it took her a year to weave the draperies.) The shadow printing and hand screening is by the McKenzie Studios of Amarillo. The fiberglass drapery next to the windows was developed by Jack Lenoir Larsen of New York.

"The carpets in the main areas were hand fabricated by V'Soske of Puerto Rico. The glass globes for the chandeliers were hand blown by Venini of Venice. . . .

"Paneling throughout the Club is dark finished oak. The Club's sunburst symbol (Edward Perrault's salute to the French king in the old club) is represented in the design of the bronze hardware throughout the 43rd floor and in inlaid woodwork on several doorways, room divider screens and cabinets.

"The floor of the main lobby is a combination of Imperial Black marble and Carnelian granite. This marble, which also is incorporated in the columns, was quarried in Tennessee. The granite is from Minnesota. The lower portion of the walls is Travertine from the Tuscany area of Italy and is the lightest color ever imported. The upper walls are paneled with padded Scottish and domestic leathers. The selection of these leathers required several months. (How did McFadden let those domestic leathers creep in there?)

"The bronze and steel sculpture on the right in the lobby was created by Russ Holmes to represent earth strata and the development of living organisms. The larger fossil specimens incorporated in this design are ammonites of the Cretaceous period collected in Texas. ('I went with McFadden to find Russ Holmes,' Upshaw would recall. 'We went to one of those artistic enclaves in Los Angeles, and McFadden knocked on the door. A woman answered. "Russ," she said, "he's out back." We went through the house and here was this man with a hood on. He was welding on something. McFadden stopped him and told him what he wanted. "You'll have to come to Houston to get my drift," McFadden said. Holmes said, "All right, Bill. I'll do it for you." That's the way it was. They all had complete confidence in McFadden, and all were ready to work for him.')

"To the right of the lobby is the reception lounge area which may be divided into several configurations by portable panels of oak and cathedral glass. In this area on two facing walls is a pair of specially designed cabinets. Above each is an intricately embroidered panel woven by Victoria Solani Baker of California, employing 50 different colors and types of yarn and cording, together with metal discs, links, nets, velvets, and linens. The striated chenille fabric was handwoven by Maria Kipp. The first panel represents 'The Origin' and depicts a blazing sun giving energy to exotic plant life of early eras.

"The second panel represents 'The Transformation,' symbolically representing the transformation of early plant life into the molecular structure of petroleum. The artist views the work 'as through a strong microscope, with the plants beginning to break down in abstract forms and the air itself filled with cellular structures.' (Victoria Solani Baker had never before been asked to weave works of such size, and at first she demurred. But succumbing to McFadden's blandishments, she finally went to work on them. When the works were finally completed and hung to McFadden's satisfaction, Victoria's husband, an engineer, pulled out his slide rule and declared that she had worked for less than the minimum wage! She said she would never again create large panels, not because of her fee, which she considered adequate, but the time and effort required was simply too much.) . . .

"The walls are oak paneled in the Petroleum Room which is 22 feet high. The dance floor is made of ebony and oak. Special strips of carpeting allow variations in the size of the dance floor. In the center of the inner wall is a huge tapestry (already described) . . .

"From the Petroleum Room we go into a corridor along the south side of the building through the service area which includes the pantry, bake shop, and kitchen. The equipment which incorporates the latest designs makes these facilities probably the most efficient of any Club facilities in the United States.

"From the service area we enter the Wildcatters Grill and Bar which, together with the card room, take up most of the west side of the Club. To our left as we enter from the service area is a card room which is separated from the grill by a removable screen.

"The ceiling over the Wildcatters Grill and Bar has been lowered to a one-story height to give a feeling of friendliness. The ceiling has nearly 1,000 tiny light bulbs integrated into a system of parallel oak fins. The upper wall surface is made of insulating cork of the type used in commercial refrigeration. Oak paneling covers the lower walls.

"The Wildcatters Bar, which is surrounded on three sides by the grill, is enclosed with specially designed phenolic resin panels. The bar top is handcarved Brazilian rosewood, and the columns are of Travertine.

"In back of the bar is a series of four metal mosaic panels designed by Wilcke Smith. On the background of oxidized copper are smaller rectangular panels of deeply toned brass and copper with designs based on the geologist's pick, seismic profiles, molecular structures of petroleum refining, and a rock bit. These four themes are repeated in numerous variations in bronze, silver, and pewter.

"Turning right at the north end of the Wildcatters Grill we go through a corridor into the library on the left and the President's Lounge on the right. This area, which will be reserved for men, is paneled in dark oak in simple modulated framing with a small brass strip at each module point. On the wall of the President's Lounge are Gittings photographs of each of our Club presidents.

"Doubling back through a part of the corridor from the Wildcatters Grill, we turn into the opening to the right and enter the Discovery Room which is a second dining area. The huge, leather-paneled, motor-driven door which separates the Wildcatters Grill from the Discovery Room is 22 feet high and 12 feet wide.

"The walls of this two-story room are of Travertine marble and padded leather. In this room is a small dance floor. The fountain in the reflecting pool is of bronze and copper and was created by Russ Holmes who describes it as 'suggesting a profile of the earth's crust shaped by recurring movement of rock layers.'

"From the Discovery Room we walk past the fountain into the cocktail lounge which extends along the north side of the building into the entrance lobby area.

"Against the inner wall is a pair of cabinets with sculptured bronze pulls in the form of trilobites which were created and cast by William Schieffer. Above the chests is a pair of metal relief panels designed by Wilcke Smith and executed in hand beaten bronze by John Wallace. Designs in the panels represent, from the lower part upward, prehistoric plant and animal forms from the Cambrian period to the Pleistocene period.

"From the cocktail lounge we proceed back into the entrance lobby and up the stairs to the 44th floor. The rooms on this floor are all available by reservation for private dining, special functions, or parties, and have dining capacities for groups from 12 to 125 people. The manager's office and Club offices are also located on this floor. In addition are two small 'quiet' rooms in which members might relax.

"The larger rooms on the 44th floor are divisible into smaller rooms with the use of wood paneled movable dividers which provide great flexibility. For these larger rooms, the entire area is called a 'suite' and is named after a producing locale. The smaller areas into which the rooms may be divided are named after producing formations within the locale. For example, the entire room on your extreme right is called the Coastal Suite, and the two rooms into which it may be divided are the Frio Room and the Miocene Room. The names of the suites and rooms are shown on the diagram included in this booklet.

"The one exception is the inner dining room on your extreme left which includes the furnishings from the Pavillion Room of our old Club and is called the Drake Room, having been named after Colonel Edwin L. Drake who drilled the first oil well in the United States. . . ."

What Pierce had done—and done superbly—was to preserve the Club's chief asset—the inspiring view of the city and beyond. Eschewing gray glass on the huge windows, which would have created a mirror affect, he utilized perimeter lighting that was slanted toward the windows to offset the brilliant outdoor sunlight. And all interior walls running parallel to the windows were in dark oak, leather or marble, while all the window walls were light marble. Any light in the Club that could be seen was purely decorative. The perimeter lighting to offset the sun's glare was recessed. The lower third of the drapes McFadden had designed did not reflect light, and a person could see through them as if they were transparent.

As for the acoustics, the cork and padded leather and the beautiful carpets were so installed that nerves frayed from drilling a dry hole would be soothed and pampered.

Gueymard had asked that the new Club be progressive and contemporary with a strong suggestion of age and maturity while reflecting the strength of character of the men who had created the domestic oil industry.

Pierce and McFadden had done that.

The old Club had been beautiful and exciting. The new Club was handsome and strong. It was Clark Gable with a touch of gray at his temples, young enough to thrill over a drilling prospect, old and wise enough to know that only the bit finds oil.

Gueymard and the members of the New Quarters Committee had been appointed by Herbert Beardmore and had served through his presidency and through the presidency of Ernest B. Miller. Miller, F. W. Bell, and Raybourne Thompson retired from the Board on April 18, 1962. They were succeeded by E. Porter Johnson, T. F. Palmer, and Walter Sterling, marking the second time Sterling was seated.

Don Connelly was elected president at the organizational meeting the next day, a fulfillment of Butch Butler's promise that Connelly would be president when the new Club would be opened. Gueymard was elected first vice-president; Amos Roberts, second vice-president; E. Porter Johnson, secretary; Wallace Thompson, treasurer; T. F. Palmer, assistant secretary-treasurer.

Since Beardmore had appointed the New Quarters Committee to serve until the new Club was ready for occupancy, it remained intact under Connelly's direction. Ernest Miller was named chairman of the Special Committee and Harry Hurt continued as chairman of the New Members Committee. F. W. Bell was appointed chairman of the House Committee; John H. Heinzerling chairman of the Entertainment Committee; Foster Parker chairman of the Finance Committee; Melvin F. Jones chairman of the Food Committee. Heinzerling was a Vinson Supply Company executive, Parker a Brown and Root executive, and Jones was a U.S. Steel division manager.

Connelly initiated a liaison system whereby the committee chairmen reported to a "Contact Director" who then reported to the Board as a whole. Amos Roberts, for example, was the "contact" between John Heinzerling of the Entertainment Committee and the Board. The system tied the Board closer to the committees and speeded up Board action on matters that seemed important to the committees.

Harry Hurt's New Members Committee moved into high gear as the time for the move approached, and the 1,100 membership quota was filled before the move was made. The club had arranged to borrow $500,000 to complete the construction and decoration of the new quarters, but needed only $325,000 of it because of Hurt's campaign. (Two months after the new Club opened the move-in bill stood at $1,837,973 with a few invoices still to come in.)

The old Club was turned over to the Ames Brothers on December 22, 1962 to allow them to benefit from the Christmas and New Year's Eve trade. Until the new Club was ready for business, the Houston Club generously opened its doors to Petroleum Club members, and the Board held its meetings in the Ramada Club. Many Club members visited the old Club, but the Ames Brothers failed to make a go of their venture.

Upshaw set up temporary offices in the Humble building, which had officially opened in September, to supervise the move. With Board approval, he staged several preliminary dining events for selected members before the opening in order to test the kitchen equipment and to acquaint the personnel with the new quarters.

The grand opening went off smoothly. Only a few in the throng were aware that a chef, who had worked 18 and 20 hours a day preparing a sugar replica of the Humble building as the centerpiece for a buffet table, had fainted from the strain.

The days and weeks that followed the opening were sometimes chaotic as members old and new flocked to the Club to show it off to friends and clients. The staff simply could not keep up with members' demands.

Walking down the corridor from the Wildcatters Grill to the lobby, Upshaw was confronted by an irate new member who had never met him. Upshaw's head was throbbing.

"I'm looking for the God damned manager of this club," said the irate new member. "Do you know where he is?"

Upshaw looked at the irate new member squarely in the eye. "I don't know where he is," Upshaw said evenly. "I'm looking for the son of a bitch myself!"

CHAPTER TWELVE

THERE HAD BEEN no preview party for the national news media as there had been back in 1951 on the Rice roof. Back then the media had been courted and the resulting publicity had warmed egos while helping to increase membership. It had seemed both necessary and desirable. Now, eleven years and two months later, the Club had matured. It had not grown sedate, but it had become an institution of sorts, an accepted part of the city's life. The opening on the Humble building roof would not mark the beginning of a new club; the Petroleum Club simply had moved to a new location. "You don't need any national publicity," James Clark, the public relations man told the Board when asked for an opinion. "Everybody who *ought* to know about the new quarters already knows."

The Grand Opening was a warm and wonderful party from the moment Martha Connelly cut the ribbon at the lobby and her proud husband signaled for the pleasantries to begin. Connelly had every right to be proud. He had moved so quickly once he knew the Humble building was available that any possible opposition had no time to form. He had overcome doubters—those who feared possible Humble dominance in Club affairs—as much by the power of his personality as by his power of persuasion. He strode among the handsome, assured men present and the ladies so elegant in their finery, accepting their compliments with ingratiating Irish charm that could not completely conceal his Irish cockiness. As president of the Petroleum Club of Houston, this night was a night of triumph for him, and the brethren showered him with their affection and respect.

Gueymard, too, received an oil drum full of plaudits. As one happy member told a group, "Connelly got us up here and Gueymard built us a home." (An ardent hunter, Gueymard was delighted when the membership later gave him an over-under Browning automatic shotgun as a token of appreciation.)

Outside the night was clear, and from the high windows the growing city sparkled as if in celebration of the club's opening. From what seemed far away, the red sign of the Rice Hotel twinkled. "There's where we came from," a member told a friend.

That faraway twinkling light had a special significance for five members present—Wilbur Ginther, Howard Warren, Vernon Frost, A. W. Waddill, and Ford Hubbard, the remaining founding fathers. Lord, how they had hoped and prayed for a club on the Rice roof!

"I couldn't stay away from the windows," Ginther would recall two decades later. "I kept looking at that Rice Hotel sign and remembering how Howard and I sat in our office in the Esperson Building and talked about starting a club for oilmen . . . . and how much we wanted it to be on the Rice roof. Howard was standing beside me, and he must have been thinking along the same lines because he said, 'It took us a long time to get up there, didn't it?' Then he said, 'We didn't outgrow that place, Wilbur. The town just grew away from it.'"

(The town would grow away from the Rice Hotel itself until eventually its door would be shut. It would sit there at what was once the city's heart like a stubborn old woman perched on her front porch in defiance of the panting bulldozers at the fence line.)

The old-timers on hand also were savoring another pleasure— the majors at long last were recognizing and supporting the club. ("Everbody but Texaco," Connelly said, "and I guess they never will.") He was right, but now the Club truly was becoming the Petroleum Club of Houston. It would draw strength from the majors, and in turn it would serve to bring the majors and the independents closer together.

Connelly, of course, presided over the first annual membership meeting in the new quarters. The meeting was held on May 28, 1963; a change in the by-laws had moved the annual meeting date from the third Tuesday of April to the fourth Tuesday of May.

Connelly thanked the membership and the board of directors for their interest and support in the past year. He said attendance and use of the Club had exceeded "our fondest hopes." In fact, Connelly said, "we have in this short time served almost enough people to fill the Rice football stadium!"

The terms of Connelly, Gueymard, Hurt, and Waddill were over, and the membership replaced them on the Board with Ivan G. Burrell, John C. Johnston, Gaston C. Jones, and Foster Parker. Burrell was a Marathon Oil Company division manager, Johnston was a Pan American Petroleum vice-president, Jones was vice-president of F. A. Callery, Inc. (oil), and Parker was vice-president and treasurer for Brown and Root.

At the organizational meeting the next day Amos Roberts was elected president. Vernon Frost was elected first vice-president and Charles Alcorn was elected second vice-president. E. Porter Johnson was reelected secretary, Foster Parker was named treasurer, and Ivan Burrell was elected assistant secretary-treasurer.

A vacancy on the Board had been created by the resignation of T. F. Palmer, who had been transferred to Tulsa. A motion was made by E. Porter Johnson and seconded by Wallace C. Thompson that Aubrey Stautberg be appointed to complete Palmer's term. The motion carried unanimously. Stautberg was president of Lago Petroleum Company.

Amos Roberts, a Baroid executive, was the first Club president to come from the "supply" category of the membership. He had begun his career with Baroid as a "mud engineer" in Ada, Oklahoma in the nearby Fitz field, at that time—1935—the most active field in the state.

Like other Club presidents, he came from a family of oil pioneers, though he kissed off the industry for a while. Eight months after "Colonel" Drake drilled his famous well at Titusville, Pennsylvania, the Rathbone brothers drilled a well in West Virginia that flowed more than 100 barrels of good oil per day from 303 feet. The well came in near the mouth of a small stream called Burning Springs Run in Wirt County.

A second well came in flowing 50 barrels per day, and the boomers flocked to Burning Springs Run. The field grew, and a village of several thousand inhabitants sprang up nearby. But the boom went flat in 1863 when General William E. Jones and 3,000 Confederate cavalrymen rode into camp. When they rode out they left derricks, boats, barrels, tanks and 300,000 barrels of oil ablaze.

Some time after the war, Amos Roberts' grandfather, Colonel D. A. Roberts, a Philadelphia and Baltimore entrepreneur, rode into the historic field and bought out the Rathbone brothers' interests. (A descendant of these Rathbones, Monroe Jackson (Jack) Rathbone, would become chief executive of Jersey Standard in the 1960s. And Amos Roberts, back in West Virginia on a visit in 1942, would find a well that was drilled in 1862 and still producing!)

All of Colonel Roberts' sons but one followed him into the oil business. The one, Amos Roberts' father, became a country doctor. And when Amos was still in grade school, the doctor moved the family to Nowata, Oklahoma. Amos, wanting to be a doctor also, enrolled at Oklahoma A & M, now Oklahoma State, as a premed student. But the desire faded; his chief interest was in chemistry, and he was graduated with a Bachelor of Science Degree in 1929.

He got a job with Texaco in a chemical plant in Tulsa where, among other things, epsom salts was manufactured. He left there to work in a Barnsdall Oil refinery, and then to work for Baroid as a mud engineer at Ada. After that he worked for Baroid in Oklahoma City, Wyoming, Canada, Illinois, then back to Tulsa as Division Manager. He was a major Baroid executive when he was transferred to Houston in 1951, and he promptly applied for membership in the Petroleum Club and was accepted.

He had worked to open the old Club; he had worked to open the new one. At the time of his presidency he was Baroid's marketing manager, world-wide.

Roberts was a fine-looking man, convivial by nature and by the demands of his trade, polite by instinct and upbringing. His wife, Dorothy Nell, was a gregarious hostess. Roberts, it will be recalled, had enlisted her aid when he was Entertainment Committee chairman during the trying days before the move. Wives of other committee members also had joined the cause of creating exciting entertainment and thereby increasing Club revenue. Other Entertainment Committees had followed suit by making the wives of members co-partners. A trend ultimately became a tradition.

John H. Heinzerling, a Vinson Supply executive, had done an outstanding job as Entertainment Committee chairman in the Connelly administration, and Roberts asked him to serve again. Walter A. Plumhoff was named chairman of the Finance Committee, Henry Mudd chairman of the Food Committee, John Domercq, Jr. chairman of the House Committee. Plumhoff was affiliated with Arthur Andersen and Company, Domercq was executive vice-president of Brewster and Bartle, and Mudd was a telephone company executive.

Roberts retained Gueymard as chairman of an abbreviated New Quarters Committee. He also appointed a Special Committee of Ernest Miller, George Barrow, and Walter Sterling. Roberts wanted this committee to dispose of the still unfinished business with Fred Heyne, Sr. and Houston Endowment. The Club had been making monthly payments to Houston Endowment since the move. A number of active Club members still believed that Heyne should forgive the Club its indebtedness by writing it off as a civic gesture. Roberts had reservations about Heyne's generosity, but he bade the committee to do its best.

Miller, Barrow, Sterling, and Upshaw met with Heyne. Miller was one of those shrewdly effective men with a fine sense of humor. He immediately set the tone of the meeting by recalling to Heyne, a devout member of the First Methodist Church, that W. F. Bryan, the church's assistant pastor, was Miller's father-in-law. And Miller also waxed eloquent about the many virtues of the church's respected leader, Dr. Paul W. Quillian.

Heyne beamed.

Thereafter, when the slightest tension developed, Miller was quick to harken back to that splendid institution, the First Methodist Church, and to its splendid leader. Meanwhile, he managed to get across his message without being forced to lay it out in flat terms. A fine and decent landlord, he seemed to say, would not be harsh with good people who had been such good tenants.

Heyne was not unappreciative of Miller's effort. But he had his answer ready. Houston Endowment, he said, was an "eleemosynary institution," relieved of paying federal taxes because its proceeds were destined for charitable grants. "I hardly think," said Heyne, "that the federal government would look upon a group of rich oilmen as suitable subjects for a charitable contribution." Houston Endowment, he said, was prepared to stick by its contract with the Petroleum Club, and Houston Endowment expected the Petroleum Club to do likewise.

Thus ended the first session.

The committee next suggested, by mail, that the debt be wiped out by the Club paying a year's rent in a lump sum. Heyne turned down the proposal. And George Barrow told the Board that it was his opinion that Houston Endowment would sue for specific performance if the Club released the old quarters back to the Foundation and gave notice of its intention to cancel the lease.

The committee tried again to settle the matter with a lump sum payment of $60,000, and again Heyne would have none of it. The Board asked Sterling to "report to Mr. Heyne that the Board of Directors of the Petroleum Club were disappointed and surprised that the Trustees of Houston Endowment were disinclined to consider a compromise settlement which we consider could be of mutual benefit." While they were at it, the Directors decided that the Club should not pay for air conditioning in the old premises since the system was not being used. Sterling was asked to discuss the matter with a qualified air-conditioning engineer. And Sterling said he would talk to Paul Wise, a Club member in the real estate business, about employing a real estate agent to sub-lease the old quarters.

Then, in mid-October, 1963, Amos Roberts received a letter from Dr. William A. Cunningham, professor of chemical engineering at the University of Texas, who was in charge of arrangements for the 56th annual meeting of the American Institute of Chemical Engineers, to be held at the Rice Hotel on December 1 through the 5th. Dr. Cunningham wrote that the Rice Hotel Sales Department had included as part of the deal the use of the 18th floor—the old quarters. Just recently, however, he had discovered that it would be necessary to receive permission from the Petroleum Club if his group were to use the space, Dr. Cunningham wrote. He wanted that permission.

Roberts propounded the Club's dilemma: the space should not be made available for the Rice to rent until the Club had some assurance that a satisfactory deal could be made with Houston Endowment. On the other hand, Dr. Cunningham's predicament could not be ignored since the chemical engineers included a large number of men in the petroleum business who were Club members. What, he asked, did George Barrow think of the matter?

Barrow said it would be unwise to let the space be used because as lessee there were certain responsibilities and liabilities that the Club could be exposed to. But, he said, since the Board hadn't had satisfactory negotiations with the landlord relative to the lease, he felt that the situation might be helpful in further negotiations.

According to the Board minutes,

> "This whole matter was discussed at length, with the conclusion that Mr. Miller and Mr. Sterling would meet again with Mr. Fred Heyne, Sr. of the Houston Endowment, Inc. and discuss with him the need for arriving at a solution of our problem that would be of mutual interest, pointing out that already the Rice Hotel had used the space both with and without permission on occasions, and that they had committed the use of the space for future meetings. . . ."

Barrow and Sterling discussed the matter privately. It was possible, Barrow said, that Houston Endowment had broken the lease. It could be a good negotiating point.

Sterling went to see Heyne. He mentioned Dr. Cunningham's letter and the occasions when the old quarters had been used by the Rice with and without Club permission. His attitude was casual; he was just making small talk, his manner indicated. Then he grew serious. He pointed out that he and Heyne had been friends for years. He knew how Heyne liked to deal, he said. "We've been telling you what *we* want. Now you tell me what *you* want."

Heyne smiled his cool smile. "Thank you, Walter," he said. "I'll send Ernie Miller a letter."

Shortly thereafter the letter arrived at the Club. It produced no joy among the Directors who still maintained that Heyne should write off the Club's indebtedness or at least settle for a year's rent. Heyne wanted to scrap the lease agreement in return for the premises and a promissory note for $183,719.80 (the amount of rental for the period commencing on October 1, 1963 and ending October 31, 1966, being 37 months @ $4,965.40 per month).

He would give the club 10 years to pay off the note in monthly installments of $1,583.33 until October 31, 1966, and the same amount monthly thereafter plus 3% interest.

Both parties would have the right to seek a tenant, Heyne wrote. If one was found, all rentals paid before October 31, 1966 would be credited against the principal amount of the promissory note.

The Rice Hotel or any other Houston Endowment affiliate would have the right to use and occupy the space on a per diem basis of $161.51 ($4,965.40 divided by 30), with the money being credited against the next maturing installment of the promissory note.

Heyne wrote also that Houston Endowment would hire an engineer to compute the cost of furnishing chilled water for air conditioning purposes, and the added cost load would be credited against the note.

Someone wanted to know if Heyne wouldn't accept a non-interest-bearing note. No, he won't, said Miller and Sterling.

Well, the $165.51 daily rental was not enough to charge the Rice Hotel or any other Houston Endowment affiliate, someone groused. The rate should be high enough to encourage all parties to find a full-time tenant.

It's high enough, said Miller and Sterling. Heyne is anxious to find a full-time tenant. And if the hotel ever needs the space on a full-time basis, Heyne won't hesitate to take it over and release us from our obligation.

There was some more grumbling.

Miller said he and Sterling had hoped for a better deal, and had tried for one, but this was the best the Club was going to get.

All right, said Amos Roberts. Nothing can be gained by more delay. Let's have a motion.

Wallace Thompson moved that the proposed settlement be accepted. The motion failed to obtain unanimous approval, but passed by a majority vote.

By the time the agreement was ready to sign by both parties rent payments had reduced the principal to $178,754.40, with monthly payments of $1,489.62. Roberts wrote a letter to the membership explaining the settlement.

Among members genuinely interested in the settlement, opinion was divided between those who had had business dealings with Heyne or who had heard of his acumen, and those who had not. The former considered the settlement excellent; the latter thought Heyne had held their feet to the fire.

Heyne later told Sterling, "They're complaining about that three percent interest and the banks are paying them more than that on certificates of deposit, which is where they have their money."

At the annual membership meeting W. O. Bartle had wondered aloud if it was proper to exceed the quota of members as set out in the by-laws . . . and as had occurred. Logan Bagby, Jr. said he thought it might be in order for the President to appoint a committee to make a study of the status of the Club membership, the procedure of admitting new members, and make recommendations to the Board for any changes it may deem to be in the Club's best interest. The present by-laws, he said, had been in effect over a long period of years, and the study might reveal a need for some changes. He put his thoughts in the form of a motion, which was carried by a majority vote.

Roberts appointed such a committee, composed of Frost, E. Porter Johnson, John C. Johnston, and Ivan Burrell. They did the job. At a special meeting of the membership, the members were asked to vote on several Board resolutions amending the by-laws and the articles of incorporation. They approved a slight alteration in the eligible business classifications, raised the limit on active resident members to 1,250, and authorized an increase in the number of shares from 900 to 1,500.

Business was good and the entertainment of high order during the year. The Club served an average of 550 meals per day. The operating profit from such patronage, plus $174,657 from new membership fees, allowed the Club to pay off its building debt ahead of schedule. With the exception of a temporary working capital loan outstanding of $50,000, the Club's only indebtedness was the note to Houston Endowment. And the Club had picked up an extra $3,000 to pay on the note with occasional rental of the old place.

At year's end the resident membership stood at 1,168; non-resident membership at 416; associate membership at 83; clergy membership at 10, and honorary membership at 13. The resident membership was classified as follows: Producers, Refiners and Transporters, 578; Professional Consultants, 72; Drilling Contractors, 50; Service Companies, 80; Lease Brokers, 16; Equipment Supply Men, 167; Royalty Owners, 28; Oil and Gas Lawyers, 42; Oil and Gas Bankers, 34; Miscellaneous, 101.

The large number of members in the Miscellaneous classification, and the fact that there still were memberships available in that category, brought a word of caution from Vernon Frost, the founding father on the Board. He said that in the early days of the Club this classification was created to admit applicants for membership who might not fit into any of the other oil industry classifications but who might be outstanding citizens of the community. In the Club's effort to fill the membership quota after it had been increased, "we may not have stressed the intended requirements for this classification," Frost said. And he suggested that since some vacancies in the classification remained, that the Club adhere more closely to the original requirements.

What Frost was saying was that in the haste to add new members, the Club had been more zealous than careful. The Board agreed that his remarks had merit.

The report of John Heinzerling, chairman of the Entertainment Committee, about that first full year of Club activities fairly sung with happiness.

> "Every special event was attended by a crowd that overflowed the Petroleum Room facilities into the Discovery Room and Wildcatters Grill," he crowed.

"During the latter part of August," he reported, "we had several Family Night parties featuring Hootenanny Folk Music Singers. With the return of the members from summer vacations, our Fall program began with the special pre-game cocktail and buffet suppers in connection with the Rice football games. The fact that the new quarters are located closer to the Southwest Freeway and also that we have better parking facilities apparently has had a lot to do with increased participation by our members and their guests. The football season started with the Rice-LSU game, and the largest crowd in the history of the Club attended both before and after the game. The before-game party was highlighted by a Dixie Land band.

"During the Fall the usual ladies bridge luncheon and style shows were held; and incidentally, for the first time in the history of the Club, models were presented in bathing suits. The male turnout was tremendous.

"Special dinner parties held during the season included the Mexican Supper and a Royal Hawaiian Party, the latter being one of the outstanding affairs in the history of the Club. The success of this party was due largely to the efforts of Mrs. G. L. Temple. The music was furnished by Herb Remington and the Beachcombers, the food was genuine Royal Hawaiian Buffet; and the members' Hawaiian dress was very colorful.

"For the first time in the history of the Club continuous music and entertainment were furnished during the Christmas holidays. Our Christmas entertainment schedule started with the Fellowship Luncheon. A Family Night Christmas Party was held with a real life Santa Claus for the children; and during Christmas week and into New Year's Eve, the famous Florian ZaBach, with his $70,000 violin and orchestra, played for luncheon as well as dinner.

"With the increased membership of our Club, our Fall Golf Tournament had the largest attendance in its history. The weather was excellent and the winners were: Low Gross, Henry David with a 74; Low Net was tie between Amos Roberts and Henry David with a 71 . . .

" . . . we are looking forward to our Annual Ball honoring the President of the Club. In view of the fact that the Club facilities for dancing required that only members were permitted to attend last year, we are again returning to the Emerald Room of the Shamrock Hilton so that members may bring guests since this room will comfortably accommodate our full requirement. Joe Reichman and his orchestra will furnish the music."

The Royal Hawaiian Party to which Heinzerling alluded produced a little drama outside the Club. While the party was being planned, one of Upshaw's assistants told him that a certain vendor would give them some banana leaves free and they could be used for decoration. Fine, said Upshaw without any great enthusiasm, having learned over the years that something free is seldom worth the price.

The leaves arrived—two bunches about the size of bundles of yard trimmings one puts at the curb for the garbage man. They were not fresh, but usuable, and Upshaw used them. A short time later, on the vendor's monthly invoice, there showed up a $190 charge for banana leaves. $190! Upshaw paid the bill for everything on the invoice but the leaves.

The leaves, at $190, were on the next month's invoice. And the next, because Upshaw still refused to pay. When the vendor got a little testy about the non-payment, Upshaw quit doing business with him. And the vendor sued the Club for $190.

Upshaw went to Justice Court, and George Barrow sent along a young associate, R. F. Wheless, whom he had given an earful of advice.

The vendor told his story: the Petroleum Club had ordered banana leaves for a party, he had delivered them, and the Club had refused to pay him the $190 charged.

Now, said young Wheless, how did you arrive at that cost for a couple of bunches of leaves?

That's my price, said the vendor.

What's the going price for banana leaves in Houston, then?

I don't know.

Did you ever sell any banana leaves to anyone else in Houston?

No.

Did you every hear of anyone else ever selling banana leaves in Houston?

No.

That's all, said Wheless (now a federal judge).

Upshaw took the stand and told his story. He described the leaves and the size of the bunches.

The judge grinned. Hell, he said to Upshaw, come out to my house if you ever need any banana leaves again. I pull more than that out of my swimming pool on a weekend. Case dismissed!

A lot had been accomplished in that first full year in the new home, and the future looked bright. Down on the streets, however, violence and tragedy and powerful changes had shaken the country like a dog shakes a bone. Aftershocks of the quake would reach the top of the Humble building.

As Connelly had given way to Roberts, the terms of Roberts, Alcorn, Thompson, and Frost came to an end. The membership replaced them with Harold Burrow, Robert Moehlman, Joseph Seger, and James U. Teague.

Burrow was a Tennessee Gas Transmission executive, Seger was an Humble vice-president, Teague was president of Columbia Drilling Company, and Moehlman, as a member of the New Quarters Committee, had drawn the first design for the magnificent tapestry on the Petroleum Room wall. He had been a bear for detail while the Club was being built. "He hated guesswork," said Wilbur Ginther. "He always wanted to know something *exactly.*" And Upshaw said, "He carried a slide rule with him to erase all doubt."

The next day at the annual organizational meeting Walter Sterling was elected president of the Club. E. Porter Johnson was elected first vice-president, Ivan Burrell second vice-president, Gaston Jones secretary, Foster Parker treasurer, and Moehlman assistant secretary-treasurer.

The son of the first president of the Humble company was now president of the Club on top of the Humble building, the Petroleum Club of Houston.

CHAPTER THIRTEEN

BY THE MID-1960s the United States had leaped out of the recessions of the 1950s onto a plateau of bright prosperity. Nowhere was this economic glitter more evident than in Houston. A million people inhabited the sprawling city. The Port of Houston ranked third in the nation. The Texas Medical Center along Fannin Street had gained international renown. A great intercontinental airport had been constructed north of the city. The Harris County Domed Stadium—the Astrodome—was a construction marvel and a magnet for tourists and conventioneers.

The oil industry was booming, in part because of the impressive growth of the country's petrochemical industry, and a great petrochemical complex had sprung up along Houston's Ship Channel and environs. New service and supply companies had sprung up and old ones had expanded to meet the ever-growing demand for their talents and wares both at home and overseas. But the conflict between independents and the majors continued, with the independents still maintaining they were being denied their share of the rapidly expanding oil and gas market. Their oil wells were still being "prorated" and natural gas was selling for what they considered a fraction of its worth under hated federal regulations.

In addition to all of Houston's other splendors, the National Aeronautics and Space Administration had located its Manned

Spacecraft Center at nearby Clear Lake, Texas, producing an explosion of electronics industries and an influx of scientists in many specialized fields.

Paradoxically, among all these manifestations of growth and plenty, the country was being torn apart by racial conflict at home and its involvement in a war in a faraway land.

A President was assassinated, and less than five years later his brother would also be killed. The leader of the black movement was slain by a white rifleman. And more and more young Americans were dispatched to Southeast Asia to fight in a war few understood and, at first, few cared about. They died just the same, 57,000 of them, and when it was over the jungles erased their footsteps in Viet Nam and at home their haunting memory would reside in a bleak but fitting memorial in Washington, D.C.

Club members found much to discuss about the oil business in the 1960s, including a continuing "scandal" that rocked the East Texas field and the industry as a whole.

The scandal first came to light in the *Oil Daily*. Under pressing questions from the paper's Houston bureau chief, Jim Drummond, the chairman of the Texas Railroad Commission, William J. Murray, admitted that "a certain amount of noncompliance with regulations on drilling deviation has been uncovered" in East Texas.

What Murray meant was that thieves had been drilling directional wells from barren acreage beyond the limits of the East Texas field back into the productive Woodbine sand. Before a series of investigations had run their courses it was concluded that some 400 "slant hole wells" had siphoned off a billion dollars in oil from the rightful owners. Civil penalties suits were filed against 150 firms and individuals, and 163 persons were named in criminal indictments. Dozens of oil companies filed hundreds of damage suits against the outlaws and many were settled without court action.

Some of those involved were established oilmen or in allied businesses. Some were landowners. Two were judges. Helping them had been an army of crooked drillers, engineers, and Railroad Commission personnel.

No one went to prison; juries refused to convict no matter how strong and damning the evidence. The state had filed civil penalties suits asking for more than $26 million from the defendants; it obtained $1,117,000. Oil companies that had been cheated did no better. In one case Texaco sued a "deviator" for $2 million and was awarded $16,000 by a jury of the "deviator's" peers.

George Wear, a Continental Oil attorney, declared that slant-hole drilling meant "trespass, theft, bribery, and a willful, intentional violation of laws and regulations. It means an intentional underground trespass on the property of another to steal. It means the filing of false reports and the making of false affidavits. It means taking advantage of the trust of others, corruption, and bribery. . . ."

In East Texas, it apparently meant something else.

Walter Sterling, the Petroleum Club's 14th president, was born in 1901, the year of Spindletop and the beginning of the Liquid Fuel Age which would so dramatically alter the course of world history. He was born on a farm near Anahuac, Texas where his father, Ross Sterling, ran a general store. He came to Houston in 1905 when Ross Sterling started the first Humble Oil Company. By the time Walter was graduated from old Central High School, Ross Sterling and others had laid the foundation for the Humble Oil and Refining Company. Walter was sent to Cisco, Ranger, and Eastland to work under the supervision of Humble's first great geologist, Wallace Pratt. (Pratt would climb through the Humble ranks and on to membership on the Jersey Standard board of directors.)

Sterling went to the University of Texas to study geology. "I loved the geology," he would say many years later, "but I couldn't cut the mustard in English, so I switched over and got a law degree."

Back in Houston he went to work with his father. Ross Sterling had resigned from Humble and had disposed of his company stock. He was still in the oil business, and he also owned some country banks and was publisher of the *Houston Post-Dispatch,* later the *Houston Post.* And Walter went into the oil business on his own.

In the Depression, Ross Sterling, who had been elected governor of the state, lost a bid for reelection and also lost his fortune.

So did Walter Sterling. "I woke up one morning and I was a quarter of a million dollars in debt," he told an interviewer. "I'm proud to say I paid them off. It took me better than five years to do it. I didn't take bankruptcy. I told my debtors, 'Give me time and I'll pay you. If you want to crack down and try to take it, go ahead because I don't have a damned thing!' "

When the U.S. entered World War II, Sterling joined the Army Air Corps. He wanted to fly, but he was told he was too old. He reported to the commandant at Turner Field, Georgia for assignment. "We need a PX officer," said the commandant. "I don't know anything about a PX," said Sterling. "Do you know the difference between your money and the PX's money?" the commandant asked. "Yes," Sterling said. "Then you're the PX officer," said the commandant.

Sterling spent the war as PX officer at Turner Field despite all his efforts to be transferred to a combat zone. "They wouldn't let me go," he would say later, "and if I do say so myself, I ran a damned good PX."

He returned to the oil business in Houston after the war, and joined the Petroleum Club in October 1948. He was immediately dedicated to its survival. And because he believed in public service he soon afterward became a trustee of the Hermann Hospital Estate, the first of several civic projects he would devote his time to.

It is best said that he loved the Petroleum Club of Houston. After its opening on the Rice roof, he was there for lunch every day that business and civic projects allowed. He was the first member of the "Round Table," which was not round at all while on the Rice roof. It was two rectangular tables Upshaw slid together to accommodate the growing bunch that wanted to eat together and talk together. When the Club moved to the Humble building, Upshaw specially ordered a round table that was placed in a corner of the Wildcatter Grill for "Sir Walter" and his comrades. By the time he was elected president Sterling had served on various committees and twice on the board of directors.

Sterling was most conservative, in his speech, his manners, his conduct, and his convictions. It will be recalled that he was the only director to vote against the resolution permitting the ladies entry to the main dining room in the old Club. In politics he was active in the conservative faction of the Democratic party.

Sterling was courtly to the ladies, however, and considerate of those with whom he differed, in the Club and in politics. He maintained a friendly calm, for example, when several Directors were critical of the deal Sterling and Ernest Miller had made with Fred Heyne, Sr. to terminate the lease on the Rice roof.

He was slyly reticent with reporters who on occasion sought to interview him. "He only knows two words," one reporter complained—"No comment." Sterling's dialogue with one interviewer went like this:

Reporter: Are you from Houston?
Sterling: No.
Reporter: Where were you born, then?
Sterling: Chambers County.
Reporter: Chambers County, Texas?
Sterling: Yes.
Reporter: In what town?
Sterling: No town.
Reporter: On a farm, then?
Sterling: Yes.
Reporter: Near what town, if any?
Sterling: Anahuac.
Reporter (grinning): Your daddy never had to whip you for talking too much, did he?
Sterling (slight smile): No.

There was no longer need for the Special Committee and the New Quarters Committee. The Library Committee and the Publications Committee had fallen by the wayside as the years had darkened in the old Club. Now Sterling was content to operate with the House Committee, Roy R. Gardner, chairman; Finance Committee, John R. Kerby, Jr., chairman; Food Committee, Gentry Kidd, chairman; and Entertainment Committee, Carnes W. Weaver, chairman. As the most recent past president, Amos Roberts was chairman of the Advisory Committee. Gardner was an oil operator, Kerby was a Gulf Oil director, Kidd was a geologist, and Weaver was head of the Carnes W. Weaver Drilling Company.

Despite a planned cutback in outside bookings, business during the Sterling presidency continued at the same healthy level it had reached after the opening of the new quarters. Such good business had made it possible for the Directors to set money aside out of each month's operation for repairs and replacements in lieu of depreciation, which had been abandoned for tax purposes. The fund was supplied with $8,333.33 per month to produce $100,000 annually. Foster Parker, the Club treasurer, had arranged for another $80,000 in surplus revenue to be invested in 4% certificates of deposit.

In Sterling's annual report to the membership he wrote:

> "We believe that the Club's long-established practice of limiting the use of its facilities to functions where attendance is made up predominantly of Club members meets the needs and wishes of the membership. The practice also meets the requirements of the Internal Revenue Service in order for the Club to retain its exemption from Federal Income Tax. Failure to limit the use of Club facilities, as mentioned, could cause its exempt status to be revoked. Your Directors and Finance Committee are pleased that membership patronage has enabled the Club to maintain a healthy financial position. . . .

Foster Parker also was able to report that the Club's indebtedness on the old quarters had been reduced considerably during the year, with more than $12,000 of the reduction coming from renting out the premises under the agreement with Houston Endowment. (Every day, when darkness fell, Upshaw had a habit of looking out the Club's north windows at the Rice Hotel. If lights were burning in the old quarters, he would phone the Rice management the next morning to find out why. It wasn't that he didn't trust the Rice management, but. . . .)

The Entertainment Committee, chaired by Carnes Weaver, again enlisted the aid of the members' wives, and the membership responded enthusiastically to the programs they arranged.

Gentry Kidd, chairman of the Food Committee, touched on a problem that had beset the Club since its birth. While praising the food and service, Kidd reported, "No doubt our most difficult problem is maintaining a steady staff of cooks and waiters. Our turnover continues to be too high, and a practical solution has not been worked out. With the continuing growth of the City, there is simply more demand than supply. We are hopeful that as more employees qualify for the Club's employee retirement plan, the turnover problem will diminish. . . ."

New clubs and restaurants seemed to open daily in Houston and cooks and waiters hurried to man them, anxious to cash in on an early business boom. Once the flush business died, they were quick to move on to another new spot. But more than that was responsible for the Petroleum Club's employee turnover. Upshaw was regarded as a fair boss by the club's staff, but he was a demanding one. He expected much from the employees. When an employee's efforts failed to satisfy his standards, Upshaw sent him on his way. He was not a tyrant by any means, but he was the boss.

Even chefs came and went. Some left for positions they considered better, but others left because Upshaw had lost faith in their talents. A few were volatile enough to try to restrict Upshaw's movements in the kitchen. The kitchen, they felt, was their domain. When Upshaw could not persuade them from that position, he sent them packing. He respected chefs, and generally liked them, but he would not tolerate any attitude or actions he believed to be contrary to the proper operation of the Club.

Some of the employees, however, had been with the Club from its beginning. One, Joe Ramirez, who would retire as an assistant manager, had worked at the Rice Hotel as a room service waiter. He had been sent by the hotel to work in the Club, and had stayed on when the Club took over its own food and service operations. (After his retirement in 1975, Ramirez insisted on staying on the payroll as a part-time employee. Members were happy to accommodate him, and he saw his old friends almost every day in his role of hospitality captain and greeter. One of his staunchest friends was Sterling, whom he had served back in the late 1940s in the Rice Hotel where Sterling was a permanent guest.)

Another token of a year of solid accomplishment was the increase in resident membership from 1,168 to 1,206.

At the annual membership meeting on May 25, 1965, Logan Bagby, Jr. rose to his feet and said he would like to express the appreciation of the membership to "Mr. Sterling, our president, and the Board of Directors, for the excellent result of their efforts."

Sterling said "that serving as president had indeed been an honor." He said that it was gratifying to him to have had such wholehearted support from the Board. And he thanked Bagby for his "kind expression."

Sterling, E. Porter Johnson, and Stautberg retired as Directors, and the membership voted to replace them with James E. Allison, Jr., W. J. Steeger, and Carnes Weaver, who had served as Entertainment Committee chairman under Sterling. Allison was an attorney with Vinson, Elkins, Weems and Searls, and Steeger was an attorney also.

At the Board's organizational meeting the next day, John C. Johnston was elected president. Ivan Burrell was elected first vice-president; Robert Moehlman was elected second vice-president; Allison was elected secretary; Parker was reelected treasurer; Joseph Seger was elected assistant secretary; Weaver was elected assistant treasurer.

That he was called Johnny Johnston by all who knew and dealt with him was an indication of the affectionate regard in which he was held. "Smart as hell," they would say of John C. Johnston, "and the friendliest man you'll ever meet."

He was known as "the friend of the independents" because, as Pan American's manager of the large Texas-Louisiana Gulf Coast Division, he would listen to a wildcatter's reasons why Pan American should supply the wildcatter with "bottom hole" or "dry hole" money. "If your geology is good and interesting," they would say, "Johnny is apt to get in the deal."

Johnston held Bachelor and Master of Science degrees in geology from the University of Wyoming at Laramie. The university's outstanding geological department had lured Johnston west from Aurora, Illinois where he had been reared. He came from a family of modest means, and he had worked during the summers in Aurora and after high school at a variety of jobs.

Johnston's interest in geology had been aroused by one of his high school teahers, R. E. Cravens, a graduate of Hanover College. Cravens taught commercial arithmetic and geography, and often took his geography classes on field trips to share his fondness for searching for fossils. Johnston began hunting and collecting fossils on his own, and a passion to study geology was born. The University of Wyoming geology school was not only good, it was inexpensive, just right for a youngster who was going to have to work at menial tasks while he managed his university studies.

And he did not relax after he obtained his B.S. degree. For his Master's work, he lived like the trappers of old, camping out in the wild, rolling country north of Laramie while he studied and recorded his impressions of the geology of the lonely area. And all the while he steeped himself in the state's oil history.

Mountain men found in oil in Wyoming in the late 1820s. It oozed from a spring on the Little Popo Agie River in what would become Fremont County. Trappers and Indians alike found the sticky substance a balm to their sores and aches.

In the summer of 1832, Captain Benjamin Bonneville of the U.S. Army was a visitor at the annual trappers' rendezvous in the Green River Valley of Wyoming. While the trappers sold their furs to company men and drank and fought for fun, they told Bonneville of the seep on Little Popo Agie.

Bonneville, exploring the Rocky Mountain area for the War Department, found the oil seep a year later. He described his find in his official report, and in 1836 he detailed his adventures to the writer Washington Irving. Irving wrote an article about the valiant captain.

Almost half a century later an entrepreneur read Irving's story and headed west from Pennsylvania. In 1884 the Atlantic-Pacific Oil Company drilled to 300 feet near the seep and discovered both oil and gas. More wells were drilled, some to 750 feet, and the production was hauled 125 miles in iron drums by wagons and sold at stations along the Union Pacific Railroad. The field would become known as the Dallas Dome field.

Though he loved Wyoming, Johnston's first job was in Shawnee, Oklahoma with Stanolind Oil and Gas. It was his introduction to the confusing, ever-changing world of oil.

Stanolind was a production arm of Standard Oil Company of Indiana. And a strong arm it was, indeed. Much of its strength came from its acquisition in the early 1930s of the Yount-Lee Oil Company of Beaumont. Miles Frank Yount had been a revered independent operator. He had initiated and made a fortune from the second Spindletop boom of the 1920s. And he had plunked down $3,270,000 for a one-half interest in the third well drilled in the great East Texas field and the 5,000 acres on which it sat. And he had found oil elsewhere.

On Yount's death, Stanolind had bought his properties for $48 million, and that one deal would make the company billions. But because Stanolind was led by daring and resourceful men, the company went on to find production in other areas.

Stanolind later would become absorbed into Pan American Petroleum, also a producing arm of Standard of Indiana. Pan American had been the creation of Edwin L. Doheny, a California wildcatter who had brought in Mexico's first producer in May 1901, only a few months after the Spindletop gusher. As Doheny gambled and won in the oil business, he gained control of a company that owned concessions covering almost the entire floor of Lake Maracaibo, Venezuela, one of the world's greatest oil reserves. Doheny had his own tanker fleet, and sat close to the seat of power in the industry.

But the vagaries of oil, politics, and international power plays left him vulnerable, and Pan American was bought out by Standard of Indiana. Like Doheny, however, Standard of Indiana had dreamed too big, and Standard of Jersey, Humble's parent, wound up with Indiana's Latin American holdings.

Standard of Indiana retreated, and placed its domestic production under the Pan American banner, and Pan American absorbed Stanolind. (Later, Standard of Indiana would do some more shuffling and everything would drop into place as Amoco.)

So Johnston went to work for Stanolind in 1939 as a geophysical helper and geological scout. World War II came along and he spent his time in the Navy. Back home, he went back to work for Stanolind as a geologist in Jackson, Mississippi. He decided to go for himself, but shortly returned to Stanolind as a geologist in Oklahoma City, and in 1948 he was promoted to district geologist back in Jackson. The next year he was transferred to Casper, Wyoming, and in 1951 he was promoted to division exploration superintendent.

He was busy. Wyoming had undergone a great oil boom during the war. New fields were found, new producing horizons were discovered in old fields. From 1942 to 1947, 41 new fields were found in 15 different counties, and deeper drilling found oil in 23 old fields. By January 1948, Wyoming's cumulative production stood at 751 million barrels.

In 1958 Johnston was transferred to New Orleans as division exploration superintendent, and the next year he moved to Tulsa as manager of exploration. In 1961 he moved to Houston as vice-president and division manager of the Texas-Louisiana Gulf Coast division, and was named to the Pan American board of directors. It was this job he held when he was elected president of the Petroleum Club of Houston.

To help guide his administration, Johnston appointed I. W. Hoskins chairman of the Entertainment Committee; E. Rue Thomas chairman of the Food Committee; George Nye chairman of the Finance Committee; J. P. Malott chairman of the House Committee. And because it was quickly decided that a new membership directory was called for, Johnston formed a New Directory Committee with Roy R. Gardner as chairman.

Hoskins was a Magnet Cove Barium executive, Thomas was vice-president of Coastal States Gas, Nye, it will be remembered, was a First City National Bank executive, Malott was a Continental Oil executive, and Gardner was an independent producer and operator.

The administration's first task was a pleasant one. The large celestial globe that had attracted so much attention and comment in the old Club had been loaned to the Museum of Natural Science in Hermann Park. It had become so popular with museum visitors that Director Teague made a motion that it be given to the museum. The motion was seconded by Ivan Burrell and unanimously carried.

In the beginning the Board was troubled by declining income. It was clear that increased membership was keeping the Club in the black because Upshaw was turning down outside business in order to adhere to IRS regulations. In one month alone he turned aside $4,000 of such business.

But the Directors did not waver. And business began picking up as the summer waned and members returned from vacations. And when the Federal Excise Tax on club dues and fees was discontinued, the membership voted to continue paying the tax as dues and fees, thus increasing the dues and fees income to the club by 20 percent.

During this term Harold Burrow resigned from the Board, and Joseph Seger vacated his seat when he was transferred to Tulsa. Burrow was replaced by Walter A. Plumhoff and C. W. Bohmer, Jr. replaced Seger. Plumhoff was affiliated with Arthur Andersen and Company, and Bohmer was in Humble's public relations department.

Upshaw had made another survey of wage scales in other clubs in the city. As a result, he recommended salary increases for 17 employees and the Board promptly approved them. Other operating costs were increasing under pressure from the inflation that was disturbing the country's economy, but Club business continued to improve and the Club's financial health was sound.

Still concerned about employee turnover, Johnston appointed a committee headed by Ivan Burrell to once again examine the pay practices of other clubs. After a study, Burrell reported that a five percent across-the-board increase in salaries and wages would place the club in the upper bracket insofar as pay scales were concerned. His committee recommended such an increase, Burrell said. It was unanimously approved by the Board.

At the end of his term in 1960, Howard Warren had suggested that some thought be given to establishing a Senior Membership category. "When older members retire from business and leave the Club, we are deprived of the pleasure of their company and the benefit of their wisdom," Warren had said.

Finding a new home and moving into it left little time for such considerations for several years, but Johnston brought the matter to the Boards' attention. "It would keep within the membership in many instances men who have made great contributions to the success of the Club," he said. He asked Walter Sterling to determine how his Advisory Committee felt about it. The Advisory

Committee recommended that the Board present the membership with a resolution creating the new membership category.

The Board did so. The membership cast 910 votes—896 For, 14 Against. The category was open to resident members who had reached 65 and who had been Club members for at least five years. A candidate would tender his share of stock to the Club and receive the going price of $1,500. Thereafter he would pay no membership fee and only half the cost of annual dues. He would be entitled to all the privileges of the Club except the right to vote and to serve on the Board of Directors. The classification was limited to 10 percent of the resident membership quota.

It was a warm gesture, and a wise one.

At the end of Johnston's term the club had 1,238 resident members.

William Parker McFadden had come down from Dallas to inspect the Club's carpeting and other furnishings. Some of the carpeting was in bad shape, prompting McFadden to comment, "A worn carpet means good business." The Directors had sweetened the existing repair and replacement fund by $100,000, and McFadden was engaged to establish a replacement program.

Finance Committee Chairman George Nye was able to paint a bright picture about current assets, pointing out that there had been a credit for rental on the Houston Endowment note of $3,309.95 during the year, leaving an outstanding balance of $127,866.47. But he also pointed out that in a few short months interest would start accruing on the unpaid principal at 3 percent per annum, and that the Club would no longer receive any rental credit for the old quarters.

Johnston, Burrell, Jones, and Parker stepped down from the Board on May 24, 1966. The membership replaced them with R. O. Garrett, John B. Holmes, M. G. Rowe, and Tom H. Wheat.

Garrett was president of Texas Gas Exploration Corporation, Rowe was chairman of the board of Rowan Drilling Company, Inc., Wheat was secretary of Transcontinental Gas Pipe Line Corporation, and Holmes was an oil operator with his Holmes Drilling Company.

At the annual organizational meeting the next day James Teague was elected president. Moehlman and Allison were elected vice-presidents, Wheat was elected secretary, Plumhoff was elected treasurer, and Weaver was elected secretary-treasurer.

Teague appointed E. Porter Johnson chairman of the House Committee, Aubrey Stautberg chairman of the Food Committee, and E. O. Buck, the early member and ramrod, chairman of the Finance Committee. I. W. Hoskins had done an outstanding job as chairman of the Entertainment Committee during the past administration, and Teague kept him on the job. He also reappointed Roy Gardner as chairman of the New Directory Committee.

Teague's tenure would see greater membership participation in regular and special activities than in any previous year, and much of it could be credited to his wise selection of committee chairmen.

CHAPTER FOURTEEN

JAMES TEAGUE was born in Caldwell, Texas in 1909, moved to booming Freeport in 1918 when his father went to work for Freeport Sulphur, and in short order was exposed to a drilling rig in nearby West Columbia where his uncle, Red Teague, was drilling a well. Red hired James' daddy to fire the boilers on the rig.

Thereafter, James Teague spent his summers in the oil patch and the rest of the year in school in Caldwell. He was graduated in 1925 at the age of 16. He worked in the field at West Columbia for a year after graduation, then moved to Houston to enter Rice Institute. Each summer, he returned to West Columbia and the field.

He was a money-maker. In the summer between his junior and senior years at Rice—in 1929—he had set up a business repairing radios, in addition to roughnecking in the field. And he was a salesman for General Electric's new-fangled refrigerators.

The Hogg family owned some 3,000 acres at West Columbia, some of it oil-producing. One day "Mr. Will" Hogg, head of the family, summoned Teague the salesman to the family ranch. Scrubbed and nervous, Teague made his appearance. "These ice boxes I've got are too damned noisy," said Hogg. "I'll take two of the kind you're selling."

As GE's star salesman wrote down the order, Hogg asked, "How much are you going to make on this deal?" Said Teague,

"If you just buy 'em, I'll make plenty." But Hogg turned to his clerk and said, "I'm going to Europe. If this boy has any need for money to finish school, you let him have it." The clerk made a note of Hogg's instructions.

Teague had never met Hogg before and the sales session lasted no more than five minutes, yet this good man believed he saw something worthy in the young salesman and, without fanfare, decided to give him help if it was needed.

Hogg died on the European trip. Teague's 20 percent commission on the $840 sale, plus what he earned in the field and repairing radios, assured him of more than enough money for his senior year. Many years later, at a time when Teague was working for the Hogg family, the clerk went to his files and found the note containing "Mr. Will's" instructions.

In 1930, when he was 20 and a Rice graduate, Teague went to work for Humble as a student engineer, but it was some time before he used a slide rule. He worked on wildcat wells in the Liberty-Beaumont area, and finally was given an engineering job near Moss Bluff. He had a company car to drive, and felt sure he was on the way to the top.

And then Humble sent him to McFadden Beach, and a marsh south of Port Arthur. Camp was in a deserted dancehall. He slept on boards between two chairs and fought mosquitos. The toolpusher put everybody to work digging a hole for a tank where no hole was necessary, as far as the young engineer and the rest of the hands could see.

So Teague quit. Quitting Humble was almost unthinkable. Quitting Humble in the middle of the Depression was a mistake of the first order.

For three months Teague sold pumping equipment—or tried to—in west Oklahoma and the Texas Panhandle. Crude was selling for 25 cents a barrel. Those with money nursed it. Teague's humorous sales pitch to potential customers: "You can lose less money using my stuff than you can by using somebody else's."

One evening in Pampa he toted up his day's work. He had made the company $1.25. A bellhop delivered his laundry. The laundry cost $1.50.

"I crawled on my knees back to Humble," Teague would say many years later.

He was taken back. He was sent to Raccoon Bend field near Belleville and, after a time, was brought into the Houston office as division engineer for the Gulf Coast division. He was 26, the youngest man in such a job, and he was making $300 a month.

He stayed with Humble until 1940, but left when "Captain Mike" Hogg hired him to supervise the Hogg's West Columbia production at twice his Humble salary. World War II came along, and Teague served his time in the Navy. When he came home, "Captain Mike" had died and the other family members wanted to sell out. Teague helped negotiate a sale, then bought their oil-field equipment to form Columbia Drilling Company. (He would run the company for 30 years, then sell it, receiving $5,000 for every dollar he had put into it.)

He was going strong when he became president of the Petroleum Club, drilling, as he would explain with a grin, "for anyone who wanted to drill and had the money to pay for it." And, he would add, "I never lost a quarter."

Teague was a perceptive man with the kind of humor that allowed him to chuckle at other's foibles and would not allow him to take himself too seriously. He had honestly believed that Robert Moehlman, not himself, should have been elected president, so he kept Moehlman busy to permit him to continue to spread his wings in club activities. And Moehlman responded. "Mention it, and he can get it done," Teague said admiringly.

And he had appointed Aubrey Stautberg, long-time member and recent Director, to head the Food Committee. Stautberg knew good food, and appreciated it. He knew good wines. He had, from time to time, been critical of the Club's food and service. He had, from time to time, made clear his feelings to Upshaw.

"I figured that if I appointed Aubrey to head the Food Committee, he and Upshaw might learn to live together," Teague said. (At the end of Teague's term, Stautberg would report to the membership: "Your Food Committee has been a working committee, I am proud to say. Any additions, deletions or other changes to the menu were made for your pleasure, and we are happy if you, as Club members, were pleased to any degree. Our efforts were made easy by the untiring assistance and complete cooperation of our General Manager [Upshaw], Chef, and Staff, along with the full support of the President and Board of Directors.") Teague's ploy had been successful. The food *was* better. And Stautberg and Upshaw were allies against the Philistines.

During the year Director Bohmer would tell the Board that he had been hearing complaints about the service, particularly at noon. And Upshaw would explain, for perhaps the hundredth time, that service personnel came and went regardless of the Board's attempts to keep them with pay hikes and fringe benefits. Since Club waiters, for example, were not tipped in cash but received a salary plus a 15 percent gratuity automatically added to a member's food check, the club withheld income taxes on both salary and gratuities. Some waiters liked to be tipped in cash so they could decide how much income taxes they wanted to pay. So they would move on.

Upshaw at this juncture was so desperate for good help that he had partially paid a Club head-waiter's vacation expenses in Florida in hopes the head-waiter could hire as many as a dozen waiters among the refugees from Castro's Cuba. Upshaw didn't want doctors and lawyers who were willing to work at anything in America; he wanted professional waiters who had worked in Havana night clubs and restaurants. But alas, the head-waiter learned that the good Cuban waiters had found work readily in Miami's hotels and restaurants.

A minor headache in comparison was the new Senior Membership classification. Sixty two members applied for Senior status within weeks after it was made available, and every following month saw others opting for it at a pace that upset the other classifications. The Board slowed the pace to some degree by changing the system of disposing of stock in the Club. It was resolved that widows of deceased members wishing to dispose of stock would be given priority. Next in line would be resigned members and, finally, Seniors. And the membership voted to alter membership classifications again to accommodate the changing times.

Moehlman had begun negotiating with Bill McFadden to refurbish the Club in the year past, and now E. Porter Johnson and his House Committee arranged a contract by which McFadden was employed on a continuing basis as a consultant, and the Board authorized replacing most of the 43rd floor carpeting, refurnishing the Card Room, completely redecorating the Coastal Suite, replacing the carpeting on the 44th floor, and obtaining new furnishings and decorations.

Johnson's committee had made yet another wage and salary study. This resulted in a general wage and salary increase, and a system was devised to allow incentive benefits. A plan for deferred income for Upshaw also was put into place.

At the end of the year Roy Gardner, chairman of the New Directory Committee, reported: "This committee was appointed for the purpose of publishing a new Pictorial Directory. I am happy to say that this is not a progress report, but rather an occasion which affords an opportunity to tell you how pleased we are for the acceptance the Directory has received. You have been very generous in your compliments to this committee, and we are certainly happy that you are pleased with the product. . . ."

Despite the worker turnover, the administration could point with pride to a noticeable improvement in menu planning and quality of food production, thanks to Stautberg and Upshaw. Hoskins and his Entertainment Committee, including wives, produced 20 planned activities, all well attended.

And E. O. Buck, chairman of the Finance Committee, could say succinctly, "Membership equity continues to increase while ample reserves have been accumulated to take care of all projected renovations and repairs to the Club quarters. Excess funds have been invested in certificates in local banks at interest rates exceeding carrying charges for the term debt on the old Club quarters. . . ."

At the annual membership meeting of May 23, 1967, Teague, Bohmer, Moehlman, and Plumhoff stepped down from the Board. They were replaced by Attorney George Barrow, Producer Roy Gardner, J. C. Posgate, and R. R. Rieke. Posgate was an Humble vice-president and Rieke, a long-time member and active committeeman, was a Schlumberger executive.

At the annual organizational meeting James Allison, Jr. was elected president. Carnes Weaver was elected first vice-president, W. J. Steeger second vice-president, Tom Wheat was reelected secretary, John Holmes was elected treasurer, and Barrow was made assistant secretary-treasurer.

Allison kept Stautberg on as chairman of the Food Committee. He made Moehlman, the retiring board member, chairman of the

House Committee. And he appointed two newcomers to chairmanship posts—Kline McGee to chair the Finance Committee, and J. J. (Jerry) Crowder to chair the Entertainment Committee. McGee was president of the Southern National Bank of Houston, and Crowder was production superintendent for Quintana Petroleum.

But Allison broke with tradition in naming his Advisory Committee. In the past this respected group had been made up of the club "owls," including some of the past presidents. On occasions the committee had as many as 35 members. Allison decided that the Club had aged sufficiently so that the Advisory Committee should be made up entirely of past presidents. So it was that his Advisory Committee was composed of Teague (chairman), Logan Bagby, Jr., W. O. Bartle, H. F. Beardmore, Hugh Q. Buck, J. R. Butler, D. L. Connelly, Michel T. Halbouty, John C. Johnston, Ernest B. Miller, Jr., Amos A. Roberts, Marlin Sandlin, R. E. (Bob) Smith, Walter Sterling, and Howard Warren.

Certainly this committee was capable of aiding any administration, and from then on each succeeding president would follow Allison's example.

Allison was only the second attorney to head the Club, the other being Hugh Q. Buck, charter member and legal advisor for more than two decades. Allison was an oil and gas lawyer, working directly under the revered "Judge" Elkins at what was then Vinson, Elkins, Weems, and Searls. And he was only 45. He generally was able to wrap his iron will and shrewdness in a coating of genuine courtesy, basted with just a spoonful of Texas drawl. "He can make you do what you don't want to do without it hurting," Upshaw one time observed. "And the hell of it, he turns out to be right."

Specifically, Upshaw was referring to Allison's insistence that a steam table be installed in the Wildcatters Grill. A steam table! In the Petroleum Club! Steam tables were for cafeterias. Company luncheonettes. Not the Petroleum Club! Not the haven that from its inception was to be the "finest in the land."

It won't demean the Club, Allison argued. The food will still be of top quality, the preparation superb. And it will allow us to serve many people quickly. It will help make up for the shortage of waiters and promote the general welfare of the Club. So do it, Upshaw.

The order went against every grain in Upshaw's heart and soul. He had hesitated before, when he thought the Directors were wrong about something, until the problem had resolved itself. Now Allison was not allowing him that option. Do it, Upshaw, Allison would say from time to time. Finally, Allison heard that Upshaw had plans to install a sandwich bar in the Wildcatters Grill. He went to see Upshaw. No, Allison said. No sandwich bar. A steam table. The Board has approved it. So Upshaw installed the steam table. It would have its critics at first, but it would remain as an integral part of the excellent food service.

Allison was a third generation lawyer. His grandfather had been a district judge in San Saba, Texas. His father practiced law in San Saba until 1921, when he took his family to the oil boom town of Breckenridge, Texas. When his father got off the train in Breckenridge, he was met by Sam Davis, the county attorney. So wild was the boomtown that Davis had to excuse himself while he went around a corner from the station to investigate a killing. James was born a year later, while that section of Texas was still wild and woolly. (Many years later, Davis would become district attorney and district judge in Houston.)

After high school, Allison went to the University of Texas, not to study law but accounting. His first accounting job after graduation paid $173 a month. When he finished serving his time in the Navy during World War II, Allison went back to school—to get a law degree. He was second in his graduating class. He had attractive offers, but he signed on with "Judge" Elkins as an oil and gas lawyer. He figured he could make more than $173 a month as an attorney.

The first problem of Allison's administration was the rapid increase in applications for Senior status. As of July 1, 1967 there were 87 Seniors with an additional 60 eligible as of that date—and that was 22 over the 125 quota (10 percent of the resident member limitation). Nine more would become eligible by October 1.

Tom Wheat moved that the five-year membership requirement be increased to 10 years, and that applications for transfer be considered on a seniority basis, seniority being defined as the date the applicant was approved for resident membership, and that the change become effective January 1, 1968. George Barrow seconded the motion and it was approved by unanimous vote.

To conform with the new Minimum Wage Law, the Board raised wages again for 80 percent of the club's 130 employees, with time and a half for overtime in excess of 40 hours a week. Jobs were classified so that compensation was based on merit as well as length of service. But the economy was still booming, and personnel turnovers continued.

Aubrey Stautberg, first a critic then an admirer of Upshaw, had learned they had more in common than a desire to provide the best in food and service to members. Upshaw, like Stautberg, knew fine wines and appreciated them. Stautberg told the Board he was hopeful that the Club would give more importance to its wine list. But to stock fine wines, he said, required proper storage facilities. He said he and Upshaw had discussed the matter at length and had concluded the Club needed a special cooling system for the storage area and an enlargement as well. The Board told him he could spend $1,000 to such end.

From this modest beginning would grow a beautiful wine cellar stocked to satisfy the most demanding palate, and this cellar and its tasting rooms would be supervised by Club members as jealous as Bacchus of their preserve.

On the other hand, the Board voted to discontinue Sunday brunch for lack of attendance. It is likely that Upshaw sighed with relief at this decision.

A golf tournament held the previous year had been such a success that it was decided to make it an annual affair, with money set aside to pay for trophies and other awards. Fred Heyne, Jr. had headed the committee staging the tournament—just another Petroleum Club job for the charter member. He had never lost his interest in the Club, nor his love for it.

And because the gin rummy players complained that the Card Room was too small (there were never more than six playing at at time), the space was enlarged. And a cigar stand, complete with humidor, was installed.

As the year ended, the level of business was still good, but increased operating costs had reduced the excess of income over expenses. But improvements had been paid for out of existing funds, and reserve funds were invested in certificates of deposit at the best possible rate of interest. So the Club was sound.

Stautberg's report to the membership for the Food Committee was so full of harmony as to be almost complacent. It said:

> "We, members of your Food Committee, are completely satisfied with the performance of the Club in the quality, preparation and presentation of the food during the past year. We feel a large majority of the members must agree as indicated by your patronage.

> "Part of our efforts have been directed toward a new menu, which I call to your attention. You will find our luncheon menu full and complete containing twenty-five items plus a 'Special of the Day,' changing daily.

> "Each member of this committee wishes to express his appreciation to the General Manager (Upshaw), Chef, and Staff for their enthusiastic cooperation, and to the President and Board of Directors for their full support."

This was, indeed, praise from Caesar!

On that happy note, Allison, Steeger, and Weaver stepped down from the Board, but not without Walter Sterling asking for—and getting—a rising vote of approval for the trio and the rest of Board, and Upshaw as well.

The membership elected C. Wardell Leisk, Kline McGee, and Kenneth Montague to the Board. Leisk was chairman of the board of Austral Oil; McGee, it will be recalled, was president of Southern National Bank, and Montague was president of General Crude Oil Company.

R. O. Garrett, the Texas Gas Exploration Corporation executive, was elected president to succeed Allison. Tom Wheat was named first vice-president, M. G. Rowe was named second vice-president, George Barrow was elected secretary, Kline McGee was elected treasurer, and J. C. Posgate was named assistant secretary-treasurer.

Many an early-day oilman was a petroleum engineer—or production engineer—simply because, by God, he said he was. Or his company insisted he was. Texas A & M University, for example, did not offer courses in petroleum engineering until 1928. And while the Aggie professors were well-grounded in theory, a sophomore named Michel Halbouty was selected to lecture on the basics of oil-well drilling because he had worked as a roustabout at Spindletop field.

Butch Butler was an engineering graduate from Miami University of Ohio, but he was a petroleum engineer because he had worked for a year in a refinery after graduation. And because he said he was. Hardly anyone every disagreed with Butler on an issue that mundane.

Literally scores of petroleum engineers were "graduates" of "ETO"—the East Texas oilfield. They were superior roughnecks and drillers who were driven by ambition to study producing sands and reservoir characteristics. In some cases that ambition and intellectual curiosity overcame educational handicaps and led to independent wealth or seats in corporate board rooms. If, as historians say, East Texas turned out millionaires like the mint turned out dimes, it turned out petroleum engineers on a similar conveyor belt. And the industry was the better for it.

Robert Oliver Garrett was a production engineer because Arkansas Fuel Oil Company said he was, and he believed it. Garrett came from a long line of doers. His great, great grandfather, Silas Garrett, was a member of the Virginia House of Representatives in 1812–1813. Silas moved to Bardstown, Kentucky and a new beginning, and his youngest son Peter journeyed to Illinois to some Revolutionary War land grants he had purchased from veterans. On his land Peter built an inclined wheel powered by oxen to grind grain and saw timber. He developed the land, built a brick yard, opened a grocery store. A town, Camp Point, was laid out, and future generations of Garretts would live there. And from there went Robert Oliver Garrett to the University of Illinois to study electrical engineering because he liked to tinker with radios and cars and had some mechanical aptitude.

But he left the university in his sophomore year, never to return, because he visited an uncle in Shreveport, Louisiana, and the uncle was an oil producer. He helped Garrett get a job with Arkansas Fuel Oil. Garrett was 20. He worked as a draftsman, roustabout, in a bull gang, and learned about cable tool drilling rigs and boomtown living in the Tullos-Urania field. His mechanical aptitude came to the fore as he "fixed things" with ease.

He was only 23 when Arkansas Fuel Oil sent him to Bartlesville, Oklahoma to install an air injection repressuring system on its leases. From there he went to Newark, Ohio to try the same thing. The producing sand was the Breagrit, and its outcropping provided material for rhinestones. The sand fractured and went to the wells . . . just as Garrett had told the Arkansas Fuel brass it would.

So he was a production engineer when he hit East Texas in 1931 (when he was 25), where Arkansas Fuel Oil owned 50 percent of the Lathrop #1, the third well in the field, and half of the acreage on which it sat. There were dozens just like Garrett in the giant field, and they were in charge of well testing and production. They respected their college-trained peers, and were respected in return. Garrett helped form the East Texas Engineering Association, and later became an officer in it. (He would become an active member in a number of industry groups.)

Garrett climbed the corporate ladder, making moves when he considered it wise to do so. and he arrived in Houston in 1955— when he joined the Petroleum Club—as an executive with Texas Gas Transmission. He was president of the firm when he became Club president, and later he became a firm director. He was a well-rounded man; he loved the oil and gas business and recognized its importance, but he was aware that it was not the sum total of existence; he studied history and politics, and was knowledgeable in both.

He had learned early that "if it ain't broke, don't fix it," so he kept Stautberg as chairman of the Food Committee and Jerry Crowder as chairman of the Entertainment Committee. He appointed Gueymard chairman of the House Committee. As Finance Committee chairman he selected Sequoyah (Squaw) Brown, vice-president of Humble Pipe Line Company.

Economists called the decade of the 1960s the longest sustained economic boom in the history of the capitalist world, but the boom was beginning to fray at the edges when Garrett took over the Club reins in mid-1968. Fueled by the Vietnam War, the U.S. Gross National Product doubled in the decade, for example, but prices for food and other essentials were rising by 1968, and would continue to do so.

To offset the cost pressures of food and labor, Garrett's administration raised the prices on the Club menu slightly. And membership dues were reduced by $1.50 while the liquor pool assessment was increased by $1.50 to improve the wine cellar without increasing total monthly payments by members.

Upshaw had told the Board that members were showing more interest in wines, and that wine sales alone now amounted to about $28,000 a year. He said the club should double its wine inventory. He said he would be able to add to the inventory when special wines were available and, because prices rise as wines age, he would like to buy some wines for holding. The Board, thereupon, voted the increase in the beverage pool fees.

Barrow, the Board secretary, reported that applications for resident memberships had been approved to fill existing vacancies, and that the Club, for the first time in its history, had a waiting list. Resident member classifications were juggled and compressed, and the annoying Miscellaneous classification was renamed "Other Individuals as authorized by the Board." The change, apparently, was cosmetic only.

If Stautberg's reports as Food Committee chairman had been succinct in previous administrations, his report during the Garrett administration could only be described as prolix:

> "We believe a principal objective of our Club is to provide its members with good food and good service. It is to this end that your Food Committee has directed its efforts.

> "It is gratifying to note that this past year has seen the greatest usage of our food facilities of any previous year; and while it is not our purpose to break or establish records, it is a happy thought for us that perhaps the Food Committee has had a part in adding enjoyment to your use of the Club.

> "Our daily contact with you has provided us with much constructive material which has helped in our planning. We realize our goal is yet to be reached; but with the continued interest and cooperation of the Club's staff, we think it is possible to maintain and surpass present accomplishments.

"The Committee and the staff, however, are fully aware of the need for improvement in several areas; and a continuing effort is being made to live up to the membership's high expectations.

"As you know, last year we introduced the luncheon buffet in the Wildcatters Grill, primarily as a measure to cope with a shortage of trained personnel. Unfortunately, the personnel problems still exist. However, the speed with which luncheon can be served buffet style has enabled us to handle more meals in the Grill, and the favorable acceptance to a faster service had made the buffet a desirable feature.

"Each year we revise our menus, retaining those items that are in popular demand and adding new and different items for the sake of variety and adventure in dining. We think the problem in a club is to avoid monotony, and a glance at the menu shows the attempts we have made to prevent this. One can actually dine at the Club many days without repeating the same meal. . . ."

It sounded as much like a lecture as it did a report, and no one had reason to doubt Stautberg's sincerity in his use of the phrase "adventure in dining." For him, exploring a new dish held some of the same excitement as approaching a new business deal, and he wanted his fellow members to share the experience with him.

And he did not forget the wine. "Quite a few of you attended the first of a series of wine seminars now programmed for the future," he reported. "Be on the lookout for notice of those to follow for you will enjoy these tastings along with the interesting and informative commentary. . . ."

Jerry Crowder's Entertainment Committee was entitled to toot a paper horn; for the first time the Christmas Family Night at the Club was held on two nights instead of one in order to help accommodate everyone who wanted to participate in the Christmas activities. Crowder, as Garrett remarked, liked a party and knew how to stage one.

Meanwhile, news from the oil patch during this period was both good and bad. The bad news was emanating from the Middle East and Africa. The good news was produced by two strong independents who were showing every promise of becoming big majors.

CHAPTER FIFTEEN

THE SMOLDERING HATREDS in the Middle East burst into flames on June 4, 1967 when Israel made a preemptive strike against Egypt. The other Arab states rallied to Egypt's defense by closing down their oil wells in a boycott against the western powers they claimed were supporting Israel. Britain and the U.S. were the Arabs' chief targets.

But Iran and Venezuela, both members of OPEC but not enemies of Israel, increased production to take up the slack. The Arabs not only lost the war but lost oil markets as well. And though sunken ships clogged the Suez Canal, the small tankers of an earlier time had been replaced by giants that easily brought vast shipments around the Cape. OPEC, never strong, was weakened and, apparently, was destined to remain that way. The boycott was lifted and the wells flowed once more.

All was well, it appeared.

A year after the Six-Day War, an American independent, Atlantic Richfield (Arco), found the largest oilfield on the North American continent—half again larger than the East Texas field. The discovery was on Alaska's North Slope at Prudhoe Bay, above the Arctic Circle.

208

Richfield alone had discovered Alaska's first major commercial field in 1957 on the Kenai Peninsula. It was called the Swanson River field.

As separate companies Atlantic and Richfield had moved into Iran with other independents to share in the five percent of the action the majors had agreed to give them to satisfy Getty's demand for independents' participation in the consortium.

As Arco, the new company moved more than 5 million pounds of equipment, mostly by aircraft, to begin drilling on the North Slope. The first well was a dry hole at 13,500 feet. The rig was moved onto acreage leased by Arco and Humble. With Arco as operator, the bit found oil in Prudhoe Bay State 1, and the field was confirmed by the drilling of Sag River State 1.

Less than a year later British Petroleum struck oil on the North Slope, and Alaska was soon alive with other oil seekers.

In another part of the world, a company headquartered in Bartlesville, Oklahoma made a discovery that ushered in another major oil province. The company was Phillips Petroleum, and the discovery was the giant Ekofisk field in the Norwegian sector of the North Sea.

It was not Phillips' first venture abroad. It will be recalled that the company was one of the 10 independents in the Aminoil group that joined forces with Getty to drill in the Neutral Zone between Kuwait and Saudi Arabia . . . and with great success.

Bartlesville was Oklahoma's first boomtown. Way back in 1875 a man named George Keeler dismounted at Caney River to find out why his horse wouldn't drink. There was an oil scum on the water, and nearby Keeler found a seep. It was only a short distance from where Jacob Bartle had set up a trading post a year earlier.

It took Keeler 22 years to get a well drilled at the seep. It those years more than 30 wells were drilled elsewhere in the Indian Territory, but none was a commercial producer. Keeler finally completed negotiations with Indians for drilling rights to 15 square miles around Bartlesville. He hauled a rig 75 miles by wagon to the drillsite, and Michael Cudahy, a wildcatter, drilled to 1,320 feet into what became known as the Bartlesville sand. On April 15, 1897, a shot of nitroglycerine brought oil rushing over the derrick.

Into this boomtown rode a couple of brothers—Frank and L. E. Phillips—buying and selling oil and gas leases and, as business prospered, building a bank—the Citizens Bank and Trust Company. By 1917 the brothers had done well enough to incorporate their holdings into Phillips Petroleum Company. Assets, they claimed, were $3 million. Number of employees, they claimed, was 27 (including nomadic roustabouts).

And how that company grew! By 1957, when it was rich from domestic triumphs and its foray into the Neutral Zone, it had grown faster than any of the 20 largest U.S. oil companies. And it had initiated a world-wide exploration program that was to culminate in the opening of the North Sea.

In 1959, a Jersey Standard affiliate and Royal Dutch-Shell had found gas in the North Sea off Holland at Groningen. And in 1965, British Petroleum found gas off the coast of England. Interest in the North Sea grew.

And then, in 1968, Phillips, as operator for a 10-company group, found the Cod natural gas and hydrocarbons field. And the next year Phillips found the Ekofisk field. It was a giant. Word of the discovery was flashed around the world, and the North Sea immediately was on every oil company's maps. In short order British Petroleum found the Forties field off Scotland, and soon after Jersey and Royal Dutch-Shell discovered Brent field off the Shetland Islands. American oil and service companies poured into the North Sea play, and so did those from other countries. Other fields would be found.

Britain, totally dependent on foreign oil, would eventually become self-sufficient because of the North Sea finds and, in fact, become an exporter. The Alaskan fields would increase U.S. reserves significantly, and the oil, by mandate, could be marketed in the U.S only. But both areas were "hostile environments" for drilling, production, and delivery, and neither Britain nor the U.S. would benefit from the great new finds in time to lessen the impact of a terrible punch from a revitalized OPEC.

Every year the club was lessened by the deaths of from five to ten resident members. The Board did not officially take note of their departures except on rare occasions. One occasion was the May 12, 1969 Board meeting near the end of Garrett's term. Garrett announced the death of M. A. Rutis, charter member. A memorial resolution for Rutis was adopted by the Board. It said:

> "Whereas, M. A. (Ves) Rutis, a charter member of the Petroleum Club of Houston, died in Houston on April 30, 1969, at the age of 73 years; and
>
> "Whereas, M. A. Rutis, a native of Sao Paulo, Brazil, had resided in Houston for 28 years and had been Secretary-Treasurer of Oil Field Salvage Co., Inc. of Houston, retiring on January 1, 1968; and
>
> "Whereas, M. A. Rutis was a veteran of World War I, a member of Central Presbyterian Church, and a member of numerous civic organizations and service clubs; and
>
> "Whereas, M. A. Rutis had been an active member of the Petroleum Club of Houston, having served on the House Committee 1950–1953 and 1953–1954, as Chairman of the House Committee 1954–1955, on the Advisory Committee 1959–1960, and on the Executive Committee of M. A. Rutus, et al (a group) since its beginning in June 1951, and had transferred to Senior Membership on July 1, 1966;
>
> "Now, THEREFORE, BE IT RESOLVED that the Board of Directors of the Petroleum Club of Houston laments the loss of M. A. Rutis and does hereby extend deepest sympathy to his surviving wife, Mrs. Bonita Rutis, and directs the Secretary to transmit to Mrs. Rutis a copy of this resolution."

Two weeks later Garrett, Holmes, Rowe, and Wheat stepped down as directors. The membership replaced them with Jerry Crowder, Don E. McMahon, Stautberg, and W. Harlan Taylor. It was the second election to the Board for Stautberg, and Crowder had served with distinction through two presidential terms as chairman of the Entertainment Committee. McMahon was president of Diamond M Drilling Company and Taylor was president of Taylor Exploration Company.

At the organizational meeting of May 29, 1969, George Barrow was elected club president. Posgate was elected first vice-president; Rieke, second vice-president; Stautberg, secretary; McGee, treasurer; and Crowder, assistant secretary-treasurer.

The new administration promptly adopted a resolution of appreciation for Garrett, praising his untiring efforts in behalf of the club.

At work, George Barrow had all the dignity and poise that being a top-notch oil and gas lawyer granted him. Relaxed, he looked very much like a cowhand who had just played a joke on the ranch foreman. Behind the Ciceronean exterior a mischevious sprite was lurking, waiting his turn.

Barrow was the Club's 19th president, the third practicing attorney to hold the reins. He had become a member in 1949, at the time of Bob Smith's presidency, and he had served when called upon ever since. He had been an early chairman of the Publications Committee. He had decried the folding of "Topics," the Club's early publication, and he had decried the demise of the Publications Committee itself. So one of his first acts was the creation of a new Publications Committee. He charged the committee, chaired by George Harcourt, sales manager of Halliburton, to produce a new publication to be called "Sunburst," and he called on James Clark, the Club's public relations expert, to aid in its manufacture. And he also charged Harcourt to produce a new pictorial directory.

Barrow kept Sequoyah (Squaw) Brown on the job as chairman of the Finance Committee, appointed Tom Wheat chairman of the House Committee, Clark R. Edgecomb, Jr. of Schlumberger chairman of the Entertainment Committee, and Robert F. Ball, Jr. chairman of the Food and Wine Committee. Ball was chairman of the board of Drilling Tools.

Barrow was reared in Houston and studied law at the University of Texas. He was graduated during the Depression, and his first job was with a government agency. He joined Baker and Botts, a prestigious Houston firm, before World War II, and returned to the firm after his military service. He set up his own practice in 1951, and in 1956 helped form the firm of Barrow, Bland, and Rehmet. He was with this firm when he became president of the Petroleum Club.

As we have seen, he took over as the Club's lawyer from Hugh Q. Buck in the early 1960s. He would call on Buck from time to time for consultation until Buck left the Club in 1969. Buck's departure and Barrow's assumption of the presidency in the same year was pure coincidence.

The increase in menu prices instituted during the Garrett administration failed to arrest the growing operating deficit. Barrow pointed out to the Board that no one wanted to raise food prices again, but that there had been no increase in dues since the Spring of 1952 except when the excise tax on dues and fees was eliminated in January 1966 by the government; the tax was retained as dues by the Club. If the financial picture doesn't brighten, Barrow said, we may have to increase dues payments.

The financial picture did not brighten. And it was obvious that a new round of wage increases was inevitable in the near future. So the Board took a proposal to the membership that Resident dues be increased from $19.50 per month to $24, and Non-Resident dues be increased from $1.50 per month to $7.50. The membership overwhelmingly approved the proposal, 749 to 45.

While they were talking about non-resident members, the Directors wondered if it was not time to extend the principal residence and business location requirements for such members to 50 miles from the Harris County Courthouse in place of the 30 miles from City Hall, as it was. They decided to think it over, and get back to it later. The matter would be left to others to decide.

Tom Wheat's House Committee got work underway to redesign and redecorate the 44th floor. Bill McFadden couldn't make himself available to do the design work, so George Pierce undertook both the reconstruction and decoration. Among other innovations, Upshaw was moved out of what amounted to a big closet to an office with a view.

Clark Edgecomb's Entertainment Committee continued the tradition of producing exciting events, and for the first time the past-presidents were guests of honor at a dinner that would become an annual event of some significance. At the first dinner and those that would follow, the menu was exciting, the food superb, the wines exquisite, the service beyond reproach and, in deference to the ladies, the oil patch camaraderie was cloaked in presidential dignity.

All men elected to serve as president of the Petroleum Club of Houston had been proud of the honor. Barrow certainly was, and he seemed to enjoy it also. His annual report to the membership carried pictures of the officers and directors and the committee chairmen as well. This had not been done before. His picture showed his happy face, the elfin quirk of a smile.

He wound up his presidential report to the membership thusly: "It was a rare privilege to work with such a fine group of gentlemen, and it has been a pleasant year for me personally. During this administration we bid adieu to the Soaring Sixties and we enter the Seventies with confidence that we will continue to maintain a superlative operation at Petroleum Club of Houston. . . ."

The Seventies would not soar. Not immediately. But the Club would maintain its balance through some of the bumpiest years in the industry's history.

At the annual meeting of May 26, 1970, Barrow, Gardner, Posgate, and Rieke stepped down from the Board. The members replaced them with Gueymard, Fred Heyne, Jr., Ike W. Hoskins, and Robert Mosbacher. Hoskins was a Dresser vice-president and Mosbacher was an oil operator. For Heyne, the charter member, his election marked the second time he had been selected to serve on the Board.

At the organizational meeting Kenneth Montague was elected president. Leisk was elected first vice-president; McGee, second vice-president; Stautberg, secretary; Gueymard, treasurer; Crowder, assistant secretary-treasurer.

Montague continued Sequoyah Brown and George Harcourt as chairmen of the Finance and Publications Committees, named James R. Lesch chairman of the Entertainment Committee, Lee Hill chairman of the Food Committee, and Rieke chairman of the House Committee. Lesch was senior vice-president and general manager at Hughes Tool Company and Hill was general tax counsel for Humble.

It was a measure of the esteem in which Montague was held by the industry that his peers elected him to the presidency of the Club when he had been a member only five years. He brought to the office a love of his industry and a strong desire to serve the Club.

He was born in Beaumont, Texas in 1916 and grew up in that oil country atmosphere. He went to Texas A&M and was graduated with a geological engineering degree. The Depression had not yet released its hold on America's economic throat. He applied for a job with Humble. Humble had none for him, but signed him up with another Jersey affiliate, Standard Oil of Venezuela (SOV).

Montague wound up at the Temblador field in eastern Venezuela, a remote area where the earth was prone to quiver and giant eels could produce an electrical charge strong enough to stun a bull, or so the nationals claimed. He stayed there a year. In his free time, Montague and a couple of buddies began building a fishing boat. They were just putting the finishing touches on it—in a company shed—when a noise made them lift their heads from their work. Standing there watching them was a group of executives from Caracas, and the leader of the group was Henry (Whopdown) Linam, SOV president.

Linam had arrived in Venezuela as a semi-literate driller; he had mastered both English and Spanish while working in the jungles, and had climbed the ladder to the SOV presidency in less than ten years. He was tougher and more demanding than a Marine drill sergeant. It was said that even the Venezuelan dictators were uneasy in his presence.

While Montague and his buddies waited for the lightning to strike, Linam inspected the boat. It was obvious that the boat had been made of company materials because they were the only materials available. Then Linam smiled. He admired their workmanship, he told the frightened trio, and he wanted to congratulate them for doing something constructive in their free time instead of sitting around on their butts. And then he was gone, with the other members of the group behind him.

It had been a trying experience for a 22-year-old rookie geologist. But it left a warm spot in Montague's heart for Henry Linam, called by the older Venezuelans *corazon de acero*—heart of steel.

Montague spent a year with another geologist on a field trip, sleeping in tents, living off the land, plotting the geology of a wide expanse of territory, trying to match the subsurface with the first aerial photographs taken in Venezuela. And then he spent some months "sitting" on a rank wildcat well.

Then Montague did an unusual thing for a young man with field experience in a foreign land. He came back to the States to fill some gaps in his education. He felt a need for additional English courses. He wanted to be better equipped for report writing. And he took speech courses to improve his public speaking ability. He got the schooling at Texas A&M and the University of Texas, finishing just in time to go into the Navy at the beginning of World War II.

After the war he got his job back with SOV. He rode from Beaumont to New Orleans on a crowded train. He sat on his suitcase. He was still forced to sit on his suitcase when the train left New Orleans for Miami. At a small junction in Alabama, Montague got off the train, called the SOV office in New York and announced that he had just quit.

Back in Beaumont he went to work for Sun Oil Company, first as a petroleum engineer and then as a reservoir engineer. And in 1956 Sun sent him back to Venezuela where the company and partners had made seismic studies in the middle of Lake Maracaibo and later obtained concessions. Montague was placed in charge of the drilling program. Sun built its own structures and platforms and other installations, and within three years was producing 200,000 barrels a day from the lake. From its terminal, Sun's tanker fleet brought the crude to the States.

Montague returned to Beaumont in 1958. Sun needed him as its Petroleum Engineering Counselor to organize and present cases before the Federal Power Commission and other government agencies. The additional schooling he had obtained was paying off.

He spent two years at that demanding chore. Then for four years he was assistant division manager in the producing department before being promoted to division manager in 1964. Shortly after that Sun gained control of General Crude Oil Company, a big independent, and Montague moved to Houston as General Crude's president. It was then that he joined the Petroleum Club.

The two years as Sun's Petroleum Engineering Counselor had sharpened his concern for the industry's general problems, and

Montague worked effectively in the various industry organizations that were trying to solve them.

And two perplexing membership problems immediately faced Montague and his fellow directors. With new companies moving their headquarters to Houston, the Club's resident member waiting list was growing monthly. Some on the list grew tired of the wait and withdrew their applications. Something had to be done, it seemed.

Something had to be done also in the Senior Member classification. The 125-member quota had been quickly filled. Many of the business retirees were short-time Club members. Now a number of long-time Club members were reaching retirement age. With the Senior quota filled, they were placed on a waiting list.

After long and careful deliberation, the Board decided to present the membership with two changes in the by-laws. One would increase the resident membership limit from 1,250 to 1,500. The other would remove the block that limited Senior Membership to ten percent of the quota for Resident Membership. Proxy ballots were mailed out to the membership, and a special membership meeting was called for December 28, 1970, at which time the voting would be recorded.

Shortly before the special meeting, the Board learned that a group of members, led by some of the oldtimers who sat at the Round Table, were strongly opposed to raising the resident membership limitation. Realizing that they could not muster enough support to defeat the resolution, they had launched a campaign to keep members from attending the special meeting. Without a quorum present—50 members—the meeting would have no validity.

The Board responded with a telephone campaign urging members to attend. Before anyone was seated it was obvious that a quorum was present, and the meeting went on.

Vernon Frost, the charter member, asked to be heard. In his calm, careful way, Frost said that having served on the Board for two different terms he knew from experience that a matter requiring the amending of the by-laws would not be put before the members for vote unless it had been carefully studied by the Directors. However, Frost said, undoubtedly there were those present who would appreciate hearing in more detail the reasons for a desire to increase the resident membership quota. He said he thought that the amendment calling for lifting the quota on Senior Members was self-explanatory.

Montague spoke. Generally speaking, he said, a Club regarded itself as in an enviable position when it had a waiting list of applicants. However, he said, in recent months the list had increased to where an extended wait was necessary. With more of industry's top management coming to Houston, the Board had concluded that the Club's purpose would be well served by making it possible for these men to join. The Club, he said, primarily was an oilman's club, and every effort should be made to accommodate "our own people." There was not a likelihood, he said, of the Club being overrun by an increase of new members.

He went on. "While the economy of the club is not a dominating factor, it is, nevertheless, a contributing influence," he said. The Club's finances were in good shape, he conceded, but it was a fact that for almost a year it had been running at a cash loss. "As all other industries, we, too, are having to deal with ever-increasing costs. Additional income is needed, and additional members will undoubtedly help in this direction."

He added that the situation warranted a modest adjustment in menu and service charges also, but that even with the adjustment the Club's prices would still be on a level with comparable clubs.

E. Porter Johnson, a Club stalwart since 1951 and a Round Table regular, rose to speak. He said he recalled quite well the days when the Club had as a goal the establishment of a waiting list of new applicants. The concern at having reached the goal, he said, was a reverse of the thinking that had prevailed over a long period of time.

"But what concerns me more," Johnson said, "is that at the present time the Club is not doing as good a job as it should in rendering good service. I thought that the purpose of belonging to a fine club was for the associations one was able to enjoy, and the pleasure of personal and specialized services. Regrettably, this does not exist now. The turnover in membership already has made me feel almost a stranger in my own Club. I feel that my views are shared by other members, and I don't feel that I can support the idea of increasing the resident membership."

Frost was recognized again. He said, "I respect the views of others, and especially those of Mr. Johnson, but in my view the President's reply to my inquiry was convincing and acceptable."

Sequoyah Brown then rose and moved that the amendments be approved.

But before a vote could be taken, Walter Sterling asked to speak. An increase in membership was a very important matter, Sterling said, and it was unfortunate that the special meeting was called during the busy holiday season. It would have received more attention had not most people been occupied with a multitude of other activities. Therefore, he said, he moved that the vote on the two amendments be tabled, and that the questions be brought before the members at the next regular annual meeting.

Montague said that the notice calling for the special meeting had specifically stated that its purpose was to approve or disapprove amending the bylaws. Not to approve or disapprove would constitute a failure to act in accordance with the purpose for which the special meeting was called. And he ruled that the vote on Brown's motion should proceed.

When the ballots were tallied, 595 members had voted for the amendment, 90 against. The amendment removing the 10 percent limitation on Senior Membership passed 615 to 69.

Setting aside E. Porter Johnson's complaint about service at the Club, there was something sadly touching in his statement that the turnover in membership "already has made me feel almost a stranger in my own Club. . . ."

Death, transfers, and resignations were thinning the ranks of the early members. Newcomers generally were younger men. Many were more conversant with the producing sands in foreign lands and beneath the seas than with the Frio, the Hackberry, the Wilcox, and the fabled Woodbine. They knew more about computers than they did about chain tongs. But they were no less oilmen, and they held every right to be applicants for Club membership.

So Johnson's plaintive cry, in truth, was against the passage of time.

The Board also moved into another membership area. Montague received some suggestions that more individuals be made honorary members of the Club. He appointed a committee of Heyne, Stautberg, and Leisk to examine the category. Heyne reported back that the committee had asked George Barrow for advice and, after considerable discussion, had agreed that the honorary membership category "no longer served a particular Club interest." The committee recommended that effective at the end of the year honorary memberships not be renewed.

Montague was authorized to write all the present honorary members to that effect. It was suggested that Montague offer the honorary members an opportunity to become resident members. He did. Only two took him up on the offer—Dr. Philip G. Hoffman and Dr. Emmett B. Fields, president and vice-president, respectively, of the University of Houston.

The opposition to raising the resident membership quota did not fade away. Some four months after the vote in favor of the increase, the Board was urged to initiate an amendment to the by-laws that would hike the resident membership fee from $1,800 to $2,500 and the stock transfer fee from $300 net to the club to $1,000.

The Board, recalling that the previous administration had increased monthly dues considerably, was content to let the matter die.

At the annual meeting in May 1971, Montague, Leisk, and McGee stepped down from the Board. The membership replaced them with W. Kenley Clark, Sequoyah Brown, and James R. Lesch.

At the organizational meeting Aubrey Stautberg was elected president. Don MacMahon was elected first vice-president; Jerry Crowder, second vice-president; W. Harlan Taylor, secretary; Brown, treasurer; and Gueymard, assistant secretary-treasurer.

In the early part of the century one could see billboards showing a boy dreaming of some day owning a Cadillac. It was a fine dream, very much a part of the American Dream and the free enterprise system.

For Aubrey Stautberg, growing up in the Heights area of Houston, the Cadillac of his dreams was driven by E. F. Woodward, an oilman who had struck it rich on the flanks of Spindletop Dome. A long Havana cigar in his mouth, Woodward would drive his 16-cylinder monster along Yale Street. Watching him, young Stautberg would tell himself, "That's what I want to be—an oilman."

He had seen oil rigs in Oklahoma, where he was born, before he moved to Houston where his father became chief electrical inspector for the city. After high school, he went to the University of Houston in the evening and worked during the day at Houston Land and Trust. He had plans to study geology, and at the bank, where he was a messenger, he studied the office manager's *Oil Weekly* before passing it on to its owner.

Later he was transferred to the bank's trust department, and there he was befriended by one Hunter Boyd, whose hobby—or avocation—was running down lost land. Boyd would take Stautberg to the courthouse with him at every chance, and he taught the young man to "run the records."

Stautberg stayed on at the bank, but he dropped out of school so he could look for a job in the oil business. It was 1935, and he had no luck. Then he ran into an old high school chum, Russell Neil. Neil was a landman, searching records for a geologist and others. "Quit the bank, and I'll show you the ropes," he told Stautberg. Stautberg had $500 and a 1931 Ford roadster. He left the bank and went for himself.

At first he worked for others, specializing in locating hard-to-find landowners or heirs for major oil companies. Then during some oil excitement in Brazoria County, he found three 10-acre tracts that belonged to midwesterners who had bought the land near the turn of the century and, for all intents and purposes, had forgotten about it. Stautberg found the owners, bought the land, then leased 20 acres to an oil company for $250 an acre. With $5,000 in his pocket, he bought a new car. It was a Chrysler, not yet a Cadillac.

He got involved in the South Houston oil play, buying town lots for $10 and leasing them to oil producers for $250. He leased one lot to Bill Noble and Romain Baker that he had obtained from a man who said he was a bachelor. Later the title revealed that the man was married and had two children. Embarrassed, Stautberg traced the children to California (the wife was dead) and cleared up the matter. Impressed by Stautberg's diligence, Noble and Baker gave him more and more business. He was able to rent a cubby-hole office in the Sterling Building.

While he was in Mineola, Texas on business, Stautberg got a phone call from Walter Mengden, a Superior Oil Company executive in Houston. Mengden said he had sent some Superior landmen to the Hawkins area to look at some land. He hadn't heard from them. Would Stautberg see if he could find them?

Stautberg couldn't find the Superior men. (It turned out they had somehow strayed to Longview.) In reporting his failure to Mengden, Stautberg told him the area Superior was interested in was "open" and could be leased at about $10 per acre. Mengden asked Stautberg to lease a block of several thousand acres for Superior. Stautberg did—and on the acreage Superior discovered the Pine Mills field. Superior kept him busy after that.

It was 1951. Stautberg was doing well—and he bought a 1951 Cadillac. And he joined the Petroleum Club. So did his friend Russell Neil.

The next year an independent operator drilled a well that produced 180 barrels per day near Gillock, in Galveston County. The land on which the well sat was properly leased, but a 100-acre tract offsetting the well had been leased to the operator by a "squatter." Or so Stautberg believed. He almost tore apart the Galveston County Courthouse in a search for evidence leading to the rightful owner of the land. He finally came down out of the courthouse attic, dirty as a coal miner, with a name and address on a letter. And he set out in search of one George Moates of Herman, Iowa.

Herman had a grocery store, a service station, and a post office. No one in the community had heard of George Moates. It was winter, and Stautberg waded through snow as he painstakingly examined every tombstone in the local cemetery. No George Moates. He plodded from farmhouse to farmhouse in the area. Finally, just was he was about to surrender, he found a family that had heard of George Moates. They believed he had gone to Galesburg, Illinois many years past.

Stautberg went to Galesburg. He found Moates' trail, and he also found out that Moates was dead. He had died a bachelor, but Stautberg learned that he had 17 collateral heirs—brothers and sisters. And they were scattered from coast to coast.

Stautberg tracked them down, one by one. He got an oil lease from them, and he agreed to file a lawsuit to try to recover the 100 acres from the squatter.

When he walked into the office of the operator's lawyer with the evidence in his hand, it was like a bomb exploding in a library. With a return to normalcy, a settlement was in order. Even the squatter got something in the deal. The heirs recovered their land, Stautberg got a lease on the 50 acres nearest the operator's well, and the operator got a lease on the remainder.

Stautberg drilled his acreage, and he made a well. And he quit the land business to become an oil operator, naming his new company Lago Petroleum Company. He prospered, and he was still president of the company when he became the Club's 20th president in 1971.

Stautberg appointed C. F. (Doc) Maginnis chairman of the Entertainment Committee, W. Kelly Clifford chairman of the Finance Committee, Milton Gregory chairman of the Food Committee, Harold E. Berg chairman of the House Committee, and David W. Peake chairman of the Publications Committee. Maginnis was a geologist for Stautberg's Lago Petroleum, Clifford was an accountant with Arthur Andersen, Gregory was director of purchasing for Quintana Petroleum, Berg was vice-president and general manager for Getty Oil, and Peake was with Milchem.

Stautberg wasted no time appointing a separate Wine Committee. He named as co-chairmen Dr. Thomas Shindler and Dr. Marshall Henry. Members were James Allison, Jr.; Fred Heyne, Jr.; Gregory, the Food Committee chairman; Andre Crispin, president of Crispin Company, international marketer of steel products. Stautberg also joined the committee.

A "palate" for food and drink is not acquired. It is a natural gift, like "an ear for music." And, like an ear for music, a palate must be educated if its owner is to appreciate its virtues.

Stautberg had a palate. Upshaw one time said of him: "He can take a sip of wine and break down its characteristics like a chemist analyzing some compound. Or he can take a bite of a dish and know the quantity and quality of herbs in it." The men on the Wine Committee shared this gift with Stautberg, as did Upshaw, but some were better "educated" than others at the time.

We have seen how in the service of three presidents Stautberg, as Food Committee chairman, did much to improve the Club's food while making the menus more exciting. Had he been less diplomatic, he would have been a thorn in Upshaw's backside. But Upshaw, himself a perfectionist, welcomed and, generally, enjoyed Stautberg's incessant cooperation.

It was with Stautberg that Upshaw first brought up the subject of retirement. Oil companies were moving their headquarters to Houston, and Upshaw anticipated Club growth and a greater number of special company parties. It was time, he said, to think about hiring and grooming a younger successor. He recommended that Stautberg interview Erik Worscheh, director of Food and Beverage Operation at the Shamrock Hilton. The Club's annual party honoring the president had been held regularly at the Shamrock Hilton because the Club was not large enough. Upshaw had become acquainted with Worscheh, had observed him at work, and had learned much about him. "He'll need some time to make the transition from a commercial operation to a private club," Upshaw told Stautberg. "He'll be ready to take over when it comes time for me to retire."

Stautberg met with Worscheh several times. He thought he had convinced Worscheh to accept the Club's proposal, but Worscheh decided to go to the Astroworld Hotel complex as assistant general manager.

Then, as he was leaving a party at the Astroworld one evening, Stautberg bumped into Worscheh. They talked about the Petroleum Club, and Worscheh's interest was rekindled. After subsequent meetings, Worscheh agreed to join the Petroleum Club staff.

Though Stautberg initiated these talks, he was not alone in the negotiations. He had satisfied himself that Worscheh had superior qualifications, but he called in others to share in making the final decision. Among them was Fred Heyne, Jr., also possessor of a palate, and the man Stautberg anticipated would succeed him as president.

At the Past-President's annual dinner, Upshaw was a surprise guest of honor. He was not told that the dinner was to become a celebration of his 20th anniversary at the Club until he found himself caught up in a "This Is Your Life" program. Marlin Sandlin told the story of how Upshaw was hired away from Williamsburg Inn. Acting for them all, Stautberg presented Upshaw with a beautiful gold Audenars Piquet watch. Don Conley presented him with the gold key to the Club that the architect had given Conley before the Club opening. And Upshaw's charming wife, Josephine Hines Upshaw, was not forgotten. Walter Sterling gallantly presented her with two dozen gorgeous roses and a jeweled American Flag brooch.

The Club's financial position took an upturn during Stautberg's administration because of increased patronage by the members and an increase in the number of company functions held in the various rooms. Saturday luncheons could not be justified economically and were discontinued.

Even so, it was obvious near the end of Stautberg's term that the national economy would of necessity have an adverse effect on the Club's economy.

In November 1971 Jack A. Horner, executive vice-president of Texas Commerce Bank, replaced W. Kelly Clifford as chairman of the Finance Committee. Clifford had moved to Germany.

In that same period, the Board—and the membership—was saddened by the death of Director W. Harlan Taylor, who was serving as Board secretary. The Board chose not to select anyone to replace Taylor, and Gueymard, as assistant secretary-treasurer, took over Taylor's tasks.

Two past-presidents died during the year—W. O. Bartle and John C. Johnston.

All three were remembered at the annual membership meeting of May 23, 1972.

At that meeting, Stautberg, Crowder, and McMahon stepped down from the Board. The membership replaced them with Harold E. Berg, Jack Colle, Milton Gregory, and L. A. J. (Jimmy) Monroe. All had been active in Club affairs. Berg was the Getty Oil executive, Colle was a geologist, Gregory was with Quintana Petroleum, and Monroe was president of Dixilyn Corporation.

At the annual organizational meeting of the Board of Directors the next day, Fred Heyne, Jr. was elected president. He was the fourth charter member to be so honored, having been preceded by Harris Underwood, Jr., Hugh Q. Buck, and Howard Warren.

Gueymard was elected first vice-president; Hoskins, second vice-president; Gregory, secretary; Clark, treasurer; and Monroe, assistant secretary-treasurer.

As an *aficionado* who kept books on wine on his bed table for study on sleepless nights, Heyne wanted to maintain the Wine Committee. He appointed James Allison, Jr. chairman of the committee with himself as liaison between the committee and the Board. The number of members being attracted to wine seminars at the Club had been increasing.

Heyne appointed Charles A. Blanchard chairman of the Entertainment Committee, John M. Sullivan chairman of the Finance Committee, John E. Lyons chairman of the Food Committee, G. Will Frank chairman of the House Committee, and John G. Yeager chairman of the Publications Committee. Blanchard was vice-president of Rowan International, Inc., Sullivan was with Arthur Andersen and Company, Lyons was vice-president of Magnet Cove Barium, Frank was senior vice-president of Austral Oil, and Yeager was with Humble.

This is what they said about partners Alwyn P. King, Jr. and Fred Heyne, Jr. in the oil patch: King would go to a hungry mountain lion with a proposition to improve the lion's hunting ability, and Heyne would produce the data to show they could do it.

King was a third-generation oilman. Heyne was a fourth-generation Houstonian. King, with his resourcefulness and audacity, and Heyne with his administrative ability, combined to make fortunes first as landmen and then as wildcatters and producers. They had been boyhood chums, and they fit together like the jaws on a wolf.

King had become a landman in 1937, swapping Mellie Esperson a wildcat lease for rent on an Esperson Building office. Heyne did not join him until 1946. By then he had been graduated from the University of Texas, had married the charming Marie Flack of

San Antonio, and had served in the Air Force during World War II. He finished up his military service in Wichita, Kansas. He had been lukewarm about the oil business, but Wichita was full of oilmen, and Heyne liked their style. When he came home from the war, he got with King, who also had just returned from service and was eager for action. "If you'll teach me, I'll learn," Heyne said. King said he would. Because Heyne's father was Jesse Jones' right-hand man, Heyne could have gone to work in any of Jones' various enterprises, but he wanted to go on his own, and with King.

They did well from the beginning, buying leases for Callery and Hurt and for J. S. (Mr. Jim) Abercrombie. But the story told around the oil patch was of their dealings with Hugh Roy Cullen, the millionaire wildcatter and philanthropist, and Arthur Temple, the lumber baron, who controlled more than 800,000 acres of East Texas timber land.

King went to Cullen's office, asking for only five minutes of his time. At the end of the second day, he got it. "What do you want, son?" Cullen asked. "I want some advice," King said. "Where should we go to look for a play that would interest a man like you?"

Cullen did not give him a direct answer. Instead, they spent four hours in a general conversation.

Then King went to Texarkana to see Arthur Temple. "I'd like to lease all your land," King said.

"For how much?" Temple asked.

"For nothing," said King. "You don't need lease money. You need to have your acreage explored and produced, and I have a man who will do it. Get your lawyers and hire some geologists, and you can take whatever you want out of every lease."

Temple pointed to a picture of his father on the wall behind his desk. "Son, you see that picture of my father?"

King said he did.

"If I leased everything for nothing, I'd have to turn that picture to the wall."

Undaunted, King went back to Cullen. Cullen said he would take the deal to drill and, if lucky, produce the acreage. King went back to Temple. Temple's lawyers were outraged but the geologists he had engaged told him to retain a quarter of each lease.

King sat outside while Temple took the deal to his board of directors. When he came out, he motioned for King to follow him into his office.

Without a word, Temple went behind his desk and turned his father's picture to the wall!

Little came of the deal because Cullen, for reasons of his own, turned back the acreage to Temple. But the story persisted as a classic example of the free enterprise system at work.

King and Heyne continued to work for Cullen, however, and for others. They got their first production when they drilled four good wells on a seven-acre lease on a flank of the old Hull field. And they made five wells at the Bloomington town-site play. They drilled, successfully, after that in other oil provinces, including more than 500 wells at Borger field in the Texas Panhandle.

And, not surprisingly for that time and place and business, all of this was accomplished without a written partnership agreement between them. Only a handshake.

Heyne was not a remake of his father. He had the elder Heyne's cool practicality but, though a private person, he was more gregarious than his father. Both men were iron-willed. Both could cut through the fat to get quickly to the lean of a matter. The senior was more austere, with a quiet humor. The junior delighted in sharing his laughter, which is no doubt why, as a member of the Petroleum Club, he worked so hard for its success. (He was twice president of River Oaks Country Club as well.)

Heyne's name does not appear in the records as one of the founding fathers of the Petroleum Club; he did not attend the earliest *formal* meetings when the charter was written and the treasury established. But he had attended informal meetings with the founding eleven, lending his enthusiastic support. And while the ink was still fresh on the charter, he was out with the others looking for a home. In truth, he was a founding father in fact if not in name.

Heyne had that rare quality of being a good listener. Young people sought him out to find ease for the problems that forever tear at the human heart. Perhaps only Upshaw knew that Club employees went to Heyne with their problems, not just when he was president but from the Club's earliest days. He had known some of them from the Rice roof times, and they came to him when their minds—or pocketbooks—were troubled.

Inflation was still gnawing at America's innards when the Heyne administration was seated. Heyne—and the directors—bit the bullet and resolved to again increase dues for resident members—this time to $360 a year. The membership approved the increase in an attempt to halt the growing operating deficit.

And, upon a motion by Sequoyah Brown, seconded by W. Kenley Clark, the Board resolved "that commencing immediately, one late payment penalty of ten percent (10%) will be added to balances of past-due accounts not paid by the 20th of the month following the previous month's billing; and that such ten percent (10%) penalty be added to the indebtedness to the Club from any such member."

New carpets were installed in the Club, but they were paid for from the refurbishing fund wisely set up in past years. H. E. Berg had been chairman of the House Committee in the previous administration when the large project was conceived. Now, as a Director, he was liaison between the Board and G. W. Frank's House Committee. It was decided also that new draperies, as striking as the ones that had been hanging for a decade, should be installed. This, too, was a major expenditure but, like the carpeting, it would be paid for from the refurbishing fund.

It will be recalled that in the early days there was a Library Committee whose chief function was to arrange for talks by esteemed geologists, tax experts, and others for the edification of the membership. It had been discontinued when the Club was in the process of moving. Now the *Petroleum Times of London* proposed to the directors that the publication sponsor a seminar on the North Sea, by now the "hottest" oil play on the globe.

The Board handed the job of supervising the event to C. A. Blanchard's Entertainment Committee. The Committee was as busy as a casing-stringing crew with regular entertainment plans, but it provided such a turnout for the event that it was necessary to borrow the huge Exxon Building auditorium to accommodate the guests. (Yes. Humble became Exxon USA in 1972. The building became the Exxon Building. Exxon, the international parent corporation known simply as Exxon, was headquartered in New York.)

Erik Worscheh had joined the staff as Upshaw's right-hand man with special charge of the food and beverage operation. His presence was felt almost immediately, but still there were complaints about poor service now and then. Gueymard noted to his fellow directors that the complaints came from newer and younger members of the Club. The waiters recognized the older members and didn't want to displease them, he said. It shouldn't be that way; everyone should receive the same good service. He added wryly, "When the waiters attend to the newer members as quickly as they do the older ones, we will have achieved perfection."

Heyne turned over the presidency of the Petroleum Club to W. Kenley Clark in May 1973. Newly-elected Directors were C. A. Blanchard, the Rowan International executive who had served as Heyne's Entertainment Committee chairman; Pete W. Cawthon, Jr., senior vice-president of First City National Bank; Willard M. Johnson, oil operator; Hiram I. Walker, president of Walker-Huthnance Offshore.

Officers elected to serve with Clark were James R. Lesch, first vice-president; Sequoyah Brown, second vice-president; Jack Colle, secretary; Milton Gregory, treasurer; Pete Cawthon, Jr., assistant secretary-treasurer.

The previous decade had been called the "Soaring Sixties." Historians would call the 1970s the "Decade of Disillusionment."

The weary old globe would shake, rattle, and roll in its anguish. And the oil industry, many would shout, was the cause of it all.

CHAPTER SIXTEEN

THE OIL GLUT spawned by the Suez Crisis of 1956 persisted throughout the 1960s. Few seemed aware when it began dissipating, though some alarms were sounded, and its disappearance and resultant shortages came as a shock to most of the inhabitants of the globe. It seemed incredible with new fields in production in Africa and the Middle East and great discoveries reported in Alaska and the North Sea that the world no longer was awash with oil.

There were several reasons for the predicament, but the chief one was simple: the oil companies had sadly underestimated the industrialized world's insatiable hunger for petroleum products. The companies didn't want a glut, but neither did they want a shortage. But it seemed as if supply stood by and watched while demand raced madly up the chart.

And while this was occurring, the companies and OPEC were locked in a fight over crude prices. OPEC had grown to 13 members but was still relatively impotent until King Idris of Libya was deposed in late 1969 by a *junta* of army officers led by one Colonel Muammar al-Qaddafi. Though Libya had become a member of OPEC, Qaddafi ignored his brothers and unilaterally sought an increase in oil prices from both the majors and independents. Libya was by now a major producer, and its crude was of high grade and almost sulphur-free.

Taking a leaf from the companies' book, Qaddafi refused to talk to them as a group. He picked them off in ones and twos, demanding a 40-cents-per-barrel increase. The companies offered five cents. Qaddafi replied by cutting Occidental Petroleum's production by a fourth to get everybody's attention. Occidental, almost totally dependent on Libyan crude, finally capitulated, agreeing to pay 30 cents per barrel more with an increase of two cents per year for the next five years. The rest, majors and independents alike, slowly fell into line.

Qaddafi's move also got the attention of the other members of OPEC. In a meeting at Caracas OPEC resolved to gain an increase and adjust taxes more favorably for its members. The members wanted to meet with the companies in Teheran to negotiate.

Qaddafi wouldn't wait for the Teheran meeting. He demanded another increase, this time of 50 cents per barrel, with additional emoluments.

Washington, D. C. and London were drawn into the fray. Messages flew across the world. The companies banded together, 23 strong, to confront the organization they had successfully ignored for a decade. International politics failed to resolve anything. And at Teheran, and later at Tripoli, so did the company front.

On April 2, 1971, it was agreed that the posted price for Middle East and African crude would be $3.30 per barrel with premiums increasing it to $3.45. Life of the agreement was five years. It would last for two.

Back in the U.S. production was on the wane, which added to the shortage. Independents had claimed for years that imported oil had all but destroyed the incentive to explore for oil in the promising basins of the country. In April 1972, a year after the Tripoli agreement, the majors told the Texas Railroad Commission and its counterparts that they would accept 100 percent production from U.S. fields. And another year later President Richard Nixon lifted all restrictions on imported oil.

Shortly before he lifted the import restrictions, Nixon slapped mandatory price regulations on major oil companies "to assure the American consumer an adequate supply of oil at reasonable prices . . ." The order, said his spokesmen, was designed "to prevent increasing pressure for higher crude oil and petroleum product prices from triggering inflationary price increases." Included in the controls was the wholesale price of gasoline.

Nixon had offered himself as a friend of oil and, God knows, some in the industry had supported him financially beyond the bounds of prudence. But palace intrigues consumed his attention, and he acted as if he were unaware that the industry was truly beleaguered. He stood by when the Congress reduced the Depletion Allowance in 1969 (it later was virtually abolished). And price controls he imposed were the last things the industry needed, particularly the independents. Nixon could have found a clue when the government in December 1972 offered tracts of submerged acreage in the Gulf off Louisiana for lease. Oil and gas companies paid out a total of $1.6 billion for 116 tracts—a full $500 million more than had ever been paid out in previous lease sales. Money exposed at the sale—the total of all bids, winners as well as losers—was an almost unbelievable $6.1 billion

OPEC, having won a price increase, now began demanding "participation." It was a bold move, even for an organization that had just discovered it had biceps. The members remembered what had happened in 1938 and 1951 when first Mexico and then Iran had nationalized the oil industry; the companies had boycotted the countries and their economies had been wrecked. Participation would be gradual nationalization in that OPEC wanted only 20 percent ownership of the concessions. Or so the countries said.

So a new struggle began. It was fought out against a background of rising tensions between Israel and the Arab states. The companies lost. And the countries sold their participation crude to other companies desperate for oil—and at prices higher than the posted price. This, of course, prompted a demand for fresh negotiations for a higher posted price. A meeting was set for October 8, 1973 in Vienna.

Meanwhile, the Arab states were pressuring the oil companies to warn the Nixon administration that continuing support of Israel could result in a cutback of Arab oil. Implied was an embargo and even nationalization. The companies sounded the alarm.

Always impatient, Qaddafi completely nationalized one concession, then announced he would take over 51 percent each of the rest of the companies in his country. And he said that the price of Libyan crude would be raised to $6 per barrel, almost twice the posted price. Further, he said he would cut off all exports to the U.S. if Washington continued to support Israel. His threat was echoed along the Mediterranean and the Persian Gulf. Goaded, Nixon warned the Libyans that a boycott could bring down on Libya a buyer's boycott, and he pointed to the Iranian experience of 1951. This empty response was uttered just shortly before the October 8 meeting in Vienna.

And on October 6, as the negotiators began arriving in Vienna, Egypt, Syria and Israel went to war. Russia was supplying Egypt and Syria, the U.S. was supplying Israel.

While the war raged, the OPEC negotiators settled on a posted price of about $5 a barrel. Company representatives offered a 15 percent increase to about $3.96, then raised it to about $4.30. The offer was unacceptable. The company men asked for a two-week recess.

While the companies attempted to convey to the Nixon administration the peril they envisioned, OPEC delegates met in Kuwait and decided to raise the price of crude to $5.12 a barrel. The Arab members agreed on an immediate cutback in production of five percent with more cutbacks in the offing until Israel withdrew from Arab territories occupied in the 1967 war and restored all rights to Palestinians. When Nixon reaffirmed his support of Israel, Saudi Arabia cut back production by 10 percent, then by 20 percent, and declared an embargo on all oil to the U.S. and the Netherlands.

While the world was still reeling from the shock, OPEC ministers met in Teheran and literally kicked the price of crude to $11.65 per barrel, more than doubling the posted price and almost quadrupling the posted price reached at Teheran and Tripoli in April 1971.

It was a brand new world—and not just for the oil industry. The era of cheap oil was over.

Back in 1950, when plans for the Club on the Rice roof were being discussed, there had been members who held misgivings about the grand design because they saw dark days of depression ahead. Bob Smith had not attempted to allay their fears. Instead, he had said, "I feel certain that the affairs of the Club will adjust themselves along with everything else, and in some manner we'll be able to carry on without any regrets." It had been a statement of simple faith.

Now Bob Smith was dead. He died on November 29, 1973 when the embargo and cutbacks with their resulting industry disarray and general economic upheaval had been in force less than two months. It was a time to remember Smith's reassuring remark because there was not a member of the Club—independent producer, major company executive, or a representative of an allied industry—who had a clear picture of the future.

W. Kenley Clark, the Club president, proposed a resolution expressing the feeling of the members and Board of Directors for the former president. The resolution was drawn up by Director Terrence McGreevy, an attorney who had replaced H. E. Berg on the Board. It said:

> "RESOLVED, that we the Directors of the Petroleum Club of Houston hereby take this opportunity to honor the memory of R. E. (Bob) Smith, one of the organizers of this Club, who died on November 29, 1973. Mr. Smith was one of the outstanding leaders of Houston and the oil industry, and exemplified this leadership by inspiring the organization of this Club. Mr. Smith served as President of this Club during the years 1950 to 1953, and through his dedication and financial support enabled the Club to secure its first permanent quarters in the Rice Hotel.
>
> "Mr. Smith's leadership attracted to the Club's membership many of the outstanding men of the oil and gas industry. In the years following, Mr. Smith continued to be an active member of the Club and to lend his guidance and support to the development of the Club into one of the outstanding clubs of this Country. But above all, Mr. Smith will be remembered as a warm and generous friend whom we were all fortunate to know; and
>
> "FURTHER RESOLVED, that copies of this Resolution be delivered to Mr. Smith's surviving widow and daughters as an expres-

sion of the sympathy of the membership of this Club to the Smith family in their bereavement.

Clark and the Directors carried on. Clark chose Gayle Whiddon as chairman of the Entertainment Committee; Robert L. Banks, Finance; John G. Yeager, Food; Benjamin Simmons, House; Daniel P. Whitty, Publications, and Aubrey Stautberg, Wine. Whiddon was with Hughes Tool, Banks was an investor. Yeager of Humble had been Heyne's Food chairman. Simmons was a geologist and Whitty was with Arthur Andersen and Company.

Clark was born in Crockett, Texas in 1911. His father was a brick mason. His mother was one of those proud women who was determined that every one of her six children was going to college. When the two eldest daughters finished high school, the family moved to Denton so the girls could go to a women's college. At graduation, the girls got teaching jobs in Austin—and the family moved there because Kenley was ready for college.

Everybody worked. Kenley labored 48 hours a week in a grocery store while taking a full course of studies except for a geology class which he attended at night. He had never heard of geology until the day he enrolled at the University of Texas. He asked a sophomore what he should take as a general science course. "Geology," said the sophomore.

Clark liked the course so much that he repeated it. He had passed, but he felt that he hadn't paid as much attention in class as he should have. By then he knew he wanted to be a petroleum geologist.

A man who shopped at the grocery store was a superintendent in Shell's pipeline department. He liked Clark, and he took him to Houston where he introduced him to Dow Hamm, Shell's chief geologist. Nothing happened right away, and Clark went to work for an uncle in Diboll, surveying land for a lumber company. He was living in a shack and pondering a bleak future when he got a telegram from Dow Hamm. Clark went to work washing samples in the Shell Laboratory for $84 a month. It was 1934, and he needed three hours of German and a geological field trip to get his BS degree.

Clark got the German credits by correspondence. Dow Hamm added four weeks to Clark's first vacation so he could make his field studies near New Braunfels. He got his degree in 1936.

He stayed with Shell for three years, then worked briefly as a scout for Transwestern Oil in San Antonio. Then he came back to Houston and Cecil Hagen, chief geologist for Superior Oil, gave him a job as a full-fledged geologist.

Early on, when Superior was interested in North Louisiana, Clark remarked to Wally Jayred, a vice-president, "If we're going to look after that area, we ought to have somebody up there."

Said Jayred: "Hell, if that's the way you feel, go on."

Clark went to Shreveport. He stayed there four years and the only Superior man he saw was one just passing through from Jackson, Mississippi. He would make recommendations for drilling acreage. Superior in Houston would ask if the chances were best for oil or gas. In most cases Clark would have to say gas. And Superior would say, "Forget it." And when Superior found a giant gas field in Louisiana, it was capped when the best price offered for the gas was three cents per mcf. It was not a happy time for company geologists.

Later on when Superior sold the gas to a pipeline company for 20 cents per mcf, it aroused the wrath of the Federal Power Commission. The Commission, not the pipeline company, said Superior had made $48 million too much off the gas sales. Eventually, Superior had to pay back $28 million.

Clark returned to Houston in 1949 as chief geologist. When Superior's headquarters was moved from Los Angeles to Houston, he became vice-president in charge of the geology section. He had become executive vice-president of the company when he was elected president of the Petroleum Club.

Clark and the Directors went about the Club's business despite the travail without. With the death of M. A. Rutis it was necessary to revamp the liquor procurement procedure, and they did so. The Club kitchen was leaking into the Exxon quarters beneath them, and they fixed that . . . temporarily. New menus were produced with special emphasis on salads for the ladies. Club lighting was improved and the Entertainment Committee could report record crowds at every event. And the medical, hospital, and group life insurance plan for employees was improved, producing for the first time a marked slowdown in employee turnover.

Happier waiters evidently resulted in improved service for business increased, and this, plus a slight increase in food prices, reversed the operating deficit trend of the past two years. This occurred despite the continued escalation in operating expenses attributable to inflationary pressures. (At an employees' meeting they rose to their feet and applauded when Worscheh explained the details of the improved insurance plan.)

Benjamin Simmons' House Committee, which had worked hard in employee relations while keeping the Club in good repair, also hung the new draperies in the Petroleum Room and began an inventory of Club furniture and fixtures with an evaluation of their replacement cost.

The Club was in a sound financial condition and its accouterments were in order when Clark, Brown, and Lesch stepped down from the Board at the annual meeting of May 28, 1974. The membership replaced them with Lee Hill, Alfred W. Roark, and Gayle Whiddon, who had been Clark's Entertainment Committee chairman. Hill was an Exxon attorney and Roark, a long-time Club member, was attorney for Cockrell Oil Corporation.

At the annual organizational meeting on June 4, Milton Gregory was elected president. Jack Colle was elected first vice-president, L. A. J. Monroe was elected second vice-president, Terrence McGreevy was named secretary, Lee Hill was named treasurer. Pete Cawthon, Jr. was elected assistant secretary-treasurer, but resigned in favor of Alfred Roark, when it was pointed out that he had held the seat in the previous administration. Club by-laws, it will be recalled, provided that no director could serve for two consecutive terms in the same office.

Gregory named John W. Phenicie chairman of the Entertainment Committee, Walter Plumhoff chairman of the Finance Committee, C. F. Maginnis chairman of the Food Committee, W. J. Mechura chairman of the House Committee, C. H. Taylor chairman of the Publications Committee, and Richard S. Ruiz chairman of the Wine Committee. Phenicie was vice-president and division manager for Amoco Production Company, Plumhoff was with Arthur Andersen and Company, Maginnis was with

Lago Petroleum, Mechura was an oil operator, Taylor was a real estate developer and Ruiz was a physician.

Milton Gregory was a soft-spoken, low-key man who had worked diligently for the Club since becoming a member in 1957. He had been particularly active on the Food and Wine Committees. He was the possessor of an exquisite palate which he exercised with common sense under the restraints of his Baptist faith. His natural dignity was such that his more earthy brothers at the Club respected his rectitude.

He was an eternal student, who read histories and biographies and studied archeology, visiting at times the ruins in Greece, Turkey, and Israel. He was an authority on Florentine art. But his special study was that of wines, and it was said of him that he could spin the history of a bouquet back to the soil that nourished the vine from which it came.

He was an oilman. He worked for only two men throughout his career and both were giants. One was Bob Smith, the other was Hugh Roy Cullen.

Gregory was born in 1913 in Texas City to a family of modest means and a strong Baptist orientation. He grew up and went to school in Houston. He worked his way through Baylor University in Waco; he managed a student dormitory, sold Greyhound Bus tickets, jewelry—anything on which he could get commission. Summers he would return to Houston and work for Cargill Printing and Stationery Company. He was graduated from Baylor in 1935 with a degree in economics and English and an abiding love for the old institution.

Back in Houston he got a job with Bob Smith by applying for one at Texaco. Smith's office manager called Texaco, where he had formerly worked, and told his counterpart that he needed a good man. The Texaco office manager gave him Gregory's name, and Gregory went to work in Smith's office.

He became Smith's trouble-shooter, traveling from Pierce Junction to Webb and Duval Counties and back to the East Texas field. He negotiated with drilling contractors, bought equipment, bought royalty, and placated angry landowners.

He also accompanied Smith to Houston's prize-fights and once, in the YMCA, he put on the gloves with Smith.

With World War II underway in Europe, Smith returned from a trip to Washington, D. C. to announce that he had been made Director of Civil Defense for a five-state region. "And," he said, "I've already hired my first man." The first man was Gregory. He moved to San Antonio as deputy director and later to Dallas. Color blindness kept him from a Navy assignment. So he spent the war as Smith's deputy.

After the war a friend spoke to Cullen about Gregory's abilities, and Cullen hired him. He became Cullen's director of purchasing. Materials were hard to get. So were supplies and equipment. Gregory had a knack for dealing with others and experience in dealing with the government. Whatever was needed, he got it. On more sophisticated terms, he became Cullen's trouble-shooter and, after Cullen's death, a problem-solver for the Cullen family and its holdings.

Gregory also had a knack for fund raising. He raised millions upon millions of dollars for his alma mater and for the Baylor College of Medicine. Baylor, on whose Board of Trustees he sat for 18 years, bestowed on him an honorary degree and, on another occasion, honored him as a distinguished alumnus. As the size of his purse grew, Gregory also became a donor, inspired no doubt by Cullen's philanthropies. On one occasion, for example, when Baylor's famed Browning Library became aware that the original of Elizabeth Barrett Browning's "Battle of Marathon" was to be auctioned in London, Gregory, in effect, signed the library a blank check with instructions not to be outbid. The library obtained the revered manuscript.

He was candid about his generosity. He gave his substance to help college students, whom he thought would be successful in life, and contributed nothing to those he considered misfits. This was contrary to the teachings of Christ and his disciples, but Gregory stuck to his philosophical guns. The logician in him wanted to see a return on his investment.

Gregory became interested in wines when he and his wife, Linda Chumney Gregory, a Baylor girl and gourmet cook, accompanied some relatives to the Moselle River Valley of Germany. At the village of Berncastel Cues, famous for its white wines and vinyards that sprout from almost vertical slopes, Gregory found he had a "knack" for wines. He had studied German at Baylor, and found he had a "knack" for the language. So it was that back in Houston at the Petroleum Club, when a question arose about German white wines, Gregory was able to discuss them thoroughly—and properly pronounce the nomenclature.

Later he developed an interest in other wines, particularly Burgundies. Like some of the other "winos" at the Club, he belonged to the major international wine societies and Escofier Society as well.

Six months after the death of Bob Smith, Marlin Sandlin died. Among the Directors, Sandlin was remembered particularly by three old-timers—Roark, Colle, and Johnson. They had worked with him before and during his presidency. And Upshaw grieved for the man who had hired him and made him friend. The five of them—Sandlin, Roark, Colle, Johnson, and Upshaw—had been young men together when the Club was young.

Roark offered a resolution which was adopted. It said:

"RESOLVED, that we the Board of Directors of the Petroleum Club of Houston desire to record our deep sorrow at the death on May 28, 1974 of Marlin E. Sandlin, who was a principal organizer and President of this Club.

"His rare skill, leadership and unselfishness which enhanced these attributes made a lasting contribution to further the purpose for which this Club was formed.

"To his wife and family is extended on behalf of the members of this Club our deepest sympathy and the hope that our appreciation of his service may in some measure lighten their burden of bereavement.

"FURTHER RESOLVED, that copies of this Resolution be delivered to Mr. Sandlin's family as a testimonial to him, not only as a valued member of the Petroleum Club, but also as a citizen of Houston."

The considerable achievements of the Gregory administration were accomplished during a time when the Board and the membership were preparing and executing a series of events designed to demonstrate their regard and affection for Upshaw, who retired in January 1975. Gregory was able to report to the membership thusly:

> "Each member is directly concerned about the financial condition of the Club. It is sound. Membership is stable and increasing slightly with a great deal of interest being shown by potential members. Also, as you have observed during the year, increasing attention is being paid to the service of food, varied entertainments for the members, and receptions for new members. A long-range program is now being carried forward, designed to keep the Club quarters in excellent appearance and to replace worn equipment. . . ."

At a black-tie party at which the members packed the Petroleum Room to overflowing, Gregory announced that the Club was sending Upshaw and his wife Josephine on a month-long tour of Europe. Because the Upshaws had never been to Europe, Gregory and his wife had dinner with the Upshaws a bit later and helped the Upshaws plan the trip, with emphasis on the wine countries. (Upshaw was a Burgundy man, leaning toward the whites.)

Before this membership party, the old crew from the New Quarters Committee had tossed a party for Upshaw. Gueymard showed him an empty memory book. "We're going to fill this up for you, Uppy," Gueymard said, "and that way we'll have an excuse to have another party for you and you can have the book for keeps." A year later, at a luncheon, Upshaw was given the memento-filled book.

The Board also made Upshaw an honorary member of the Club, and the Silver Anniversary Pictorial Directory was dedicated to him.

For Upshaw, it was a time of memories, most of them pleasant. For more than 23 years the Club had been a major part of his life. He could recall the time when the Club was new and Bob Smith asked him if he had a job for an aging supply man who had fallen on bad times. Upshaw hired Vernon (Red) Gresham as night auditor. Gresham was a cheerful man who seemed to know everybody in the oil business. Upshaw wondered if he wasn't wasting Gresham's talent. He made him a Club greeter.

Upshaw's office was so located that he could not help but see Gresham at his new job on the first day. Shortly before noon an elevator opened and out stepped several men. Gresham rushed to one of them, threw out his arms and cried, "Jim, you old son of a bitch, it's good to see you!" Jim, happily surprised, boomed, "By God, Red, it's good to see you, too!"

Upshaw quickly drew Gresham aside and explained that he hadn't intended for Gresham to be *that* friendly. Gresham toned down his style, but he always greeted members and guests with a genuine warmth that brightened the day. He stayed on the job until about six months after the move to the new quarters, retiring in his seventies.

Hardly a week went by that Upshaw wasn't called on for advice by an employee, and he was touched one day when a cocktail waiter said to him, "You know how we feel about you, Mr. Upshaw. We feel like you're our father." But Upshaw took advice as well. He had a strict rule that barred the wearing of beards, moustaches, and long hair. One day Alonzo Bell came to him. Bell had started out with the Club as a bus boy on the Rice roof. He would become a waiter and then a captain in the cocktail lounge and the Men's Grill. Bell suggested that Upshaw relax his rule. Bell was not there to intercede for others but for the good of the Club. "We can keep more good men that way," he explained earnestly. The rule was relaxed.

And then there were the tamales. Upshaw had selected one day of the week as "South of the Border" day when the specialty of the house was Mexican food. It was generally agreed that the tamales served were terrible. Aubrey Stautberg, who remembered vividly the taste of the tamales he bought out of a five-gallon lard can on a Heights street corner when he was a lad, conferred with Upshaw. Upshaw spoke to the chef, who had been buying the tamales from a local tamale factory. "Go to the best Mexican restaurant," Upshaw told the chef. "Get the best that can be obtained."

The chef shrugged and pointed to an elderly Mexican-American woman whose sole job was the husking of shrimp. "She can make the greatest tamales you ever tasted," said the chef. Upshaw sighed and restrained himself from throttling the chef. "You are now the tamale maker," he told the lady shrimp husker. She glowed with happiness and made out a shopping list for the chef. Topping the list was a hog's head. The chef looked at Upshaw with raised brows. "Get it," said Upshaw. So every week a hog's head was delivered to the Club kitchen.

The tamales were wonderful. So wonderful that Upshaw on the first day they were prepared ate himself sick enough to have to call a doctor! And Stautberg touted the tamales to one and all. "Have you tried the tamales yet?" he asked one member after another. One member replied, "I ate some last week, Aubrey. And you know, I found a tooth in one." Stautberg, thinking of the hog's head, said soothingly, "That was just a whole kernel of corn, my friend. I find one now and then myself."

And don't forget the catfish. Both members and the ladies liked it very much and it was on both the men's and ladies' luncheon menus. Then a rumor floated down from Washington, D. C. that Jackie Kennedy, on a trip to Texas, would be a luncheon guest at the Club. The President's lady, Upshaw thought, might be shocked by the word "catfish" on the menu. He toyed with the idea of taking it off, but realized he couldn't deprive the ladies of the delicacy. Thinking back on his youth, catching catfish from a creek bank, he got an idea. On the ladies' menu, "catfish" was changed to "willow trout." The rumor perished, but "willow trout" remained on the menu.

He had been offered jobs elsewhere over the years, and a few had been tempting, but Upshaw had turned them down. The Club enjoyed an international reputation for excellence and Upshaw, as manager, basked in the same sunlight. He had some doubts about the wisdom of accepting the Petroleum Club job at first; he and Josephine had arrived in Houston in August and the heat and humidity had been enervating. The stately trees and greenery of Virginia were longingly recalled. But the friendliness with which they were greeted and accepted blunted their homesickness and finally erased it.

Every administration had praised him, and many a president had credited him with the Club's success. "We couldn't have made it as well as we did without him," said Walter Sterling firmly. "He has the qualifications it takes to run a major oil company or any large enterprise. He is one in a million."

Said W. Kenley Clark: "He had the touch. He treated every member as if he were the president of the Club."

True, but the staff on one occasion saw a sterner Upshaw. A member walked into Upshaw's office, sat down, leaned over and drummed his fingers on Upshaw's desk. "I've got a complaint," he said, drumming the desk, glaring at Upshaw.

"Fine," said Upshaw. "Now take your hands off my desk and sit back in your chair and tell me what's on your mind."

The member lifted his fingers, sat back, and said, "I don't like the way this Club is being run."

Upshaw nodded. "If your complaint is against the employees, tell me. If it's against me, tell the Board of Directors." He waited, and the man got up and left. The Board never heard from him, and he was still a steady member when Upshaw retired, according to the staff. "Mr. Upshaw was always nice to him, and he was always nice to Mr. Upshaw," said a staff member.

So he was an unusual man.

And, fortunately for the Club, so was the man who replaced him.

Erik Worscheh was born in Czechoslovakia, served in the German Navy, served in the British Navy, worked in the American PX at Nuremberg, and got lost in a blinding rainstorm on Houston's Main Street. Along the way he had chatted with film stars, two U.S. Presidents, and gained a national reputation as a hotel man. He had always wanted to be a diplomat, and he finally found a job to suit his desire when he became general manager of the Petroleum Club of Houston.

Worscheh came from a long line of innkeepers. He worked in his father's hotel as a lad and, while attending the University of Prague, he worked the summer months at the Hotel Imperial and the Hotel Pupp in nearby Karlsbad. All the while he was studying for the diplomatic service. It came to an end when German armies occupied the country. He was swept up with other young Czechs to serve the Nazis. Most went into forced labor. Some, like

Worscheh, were thrust into the German Navy as non-volunteers. There he joined Poles, Hungarians, and other Central Europeans who had been caught in the dragnet.

Worscheh was sent to the Baltic Sea aboard a mine-sweeper to clear the Russian mine fields approaching Helsinki and Leningrad. It was dangerous work. Since he had no alternative but to serve, he served well, and was promoted to lieutenant.

When the tide of battle shifted, the German Navy retreated in the Baltic as the German Army was doing on land. In 1944, while his navy group was near Kiel in Germany, it was captured by the British. Those with Nazi party affiliation were sent to prisons. Worscheh was interrogated. The British, to his surprise, had a complete file on him. He was offered a lieutenancy in the British Navy—and he took it. Now his job was to escort Russian and British vessels through the mine fields to Russian ports.

Meanwhile, the Russians had occupied Czechoslovakia, and Worscheh could hear no word from his family—mother, father, sister, and younger brother. The Red Cross and other agencies could not help him.

Then, while on leave, he was accosted on the streets of Hamburg by a woman with shorn hair. She was a high school classmate, and she had just escaped from the Russians. She told Worscheh that his parents had escaped the Russians and were in a refugee camp in Ansbach, Germany, which was under American control. On Mother's Day in 1947, Worscheh walked into his mother's arms.

The Russians had confiscated the family property, so the Worschehs opened a restaurant in Ansbach. Erik stayed a year, then headed for the University of Nuremberg. He had put aside his thoughts of the diplomatic service; he studied restaurant-hotel management and economics. To support himself he worked in the American military PX—and he applied for immigration to the U.S. He obtained his B.A. degree and was working on his Master's thesis when his papers were approved.

He arrived in the U.S. in 1951. At the historic Mission Inn at Riverside, California he got a job as a pantryman. He worked his way up, to Banquet Manager and then Maitre d' Hotel. He moved on then to the Beverly Hilton in Beverly Hills as Assistant Room Service Manager and Banquet Captain. It was there that

Harry Truman gave him an autographed copy of his memoirs. And it was there, while an assemblage waited, that President Eisenhower took fifteen minutes to explain to Worscheh why the American forces had held back and let the Russians occupy Czechoslovakia. When an FBI agent paled because Worscheh hadn't been "cleared" to serve the President, as had the rest of the staff, Ike said, "You don't have to worry about this fellow. He's fit to ride the river with." And three weeks later Worscheh received a fine photo, properly inscribed, from the President. It was something he would cherish.

From the Beverly Hilton he went to the Statler Hilton in downtown Los Angeles to manage the Steak House. The place had been serving about 50 persons a day. After a year it was serving about 400.

And then Bob Leroy, manager of the Shamrock Hilton, asked him to come to Houston as Director of Banquet Sales. He drove from Los Angeles to Houston, arriving in a blinding rain. He drove out Main Street from downtown. Not far from the Shamrock, thinking he was too far out of town, he turned to his left and bumped up on an esplanade. His car stuck. Worscheh got out in the storm, and a passing motorist took him to the Shamrock Hilton.

It was midnight. Standing in the lobby, soaked to the skin and bone weary, Worscheh told the night clerk, "I'm Erik Worscheh, your new Director of Banquet Sales."

"Uh huh," said the clerk. "Let me see some identification."

"It's in my car."

"Where's your car?"

"Up the street on the esplanade."

The night clerk quickly called the security chief. "I've got a nut out here."

The security chief was R. R. Simmons, a former police detective with several notches on his gun. He gave Worscheh the third degree while Worscheh kept pleading with them to call Bob Leroy, the manager. Leroy lived not in the hotel but out in the city. Finally Leroy was called.

Immediately Worscheh was whisked to a suite, his car was fetched, his baggage brought up. Leroy arrived at 2 a.m. with a bottle of champagne. And at 8 a.m. Worscheh was introduced to the news media.

He was Director of Food and Beverage Operations when Upshaw recommended him and Stautberg and Heyne and others attempted to induce him to come to the Petroleum Club. And he was Assistant to the General Manager, Food and Beverage, at the Astroworld Hotel Complex when he finally succumbed to their blandishments. He had served under Upshaw from 1973 to 1975 when he took over as General Manager of the Petroleum Club of Houston.

He was tall, affable, and he knew what he was doing. And at heart he was a wildcatter.

Hiram Walker had resigned from the Board during the Gregory administration, and had been replaced by Ross Bolton, vice-president of Texas City Production & Exploration. At the annual membership meeting of May 27, 1975, Gregory, Colle, Mc-Greevy, and Monroe left the Board. To replace them the membership elected A. L. Ballard, C. F. Bowden, C. F. Maginnis, and W. Henson Moore. Ballard was president of Kilroy Company of Texas, Bowden was vice-president of Union Oil of California, Moore was president of the Offshore Company. Maginnis, it will be recalled, was an active committee chairman.

At the annual organizational meeting the next day, Pete Cawthon, Jr. was elected president of the Club. Willard Johnson was elected first vice-president; C. A. Blanchard, second vice-president; H. Ross Bolton, secretary, Alfred Roark, treasurer; and Gayle Whiddon, assistant secretary-treasurer.

Cawthon retained W. J. Mechura as chairman of the House Committee and C. H. Taylor as chairman of the Publications Committee. He named R. M. Edwards chairman of the Entertainment Committee, Edwin I. Davis chairman of the Finance Committee, J. P. Watson chairman of the Food Committee, and Dr. T. O. Shindler chairman of the Wine Committee. Edwards was with Magcobar, Davis was an accountant, and Watson was general manager of Southeastern Public Service Company.

The year ahead would be busy and productive.

CHAPTER SEVENTEEN

THE AMERICAN PEOPLE had not been so angry since Pearl Harbor. While being forced to wait in long lines to buy gasoline at rising prices during the embargo, they directed their ire not at the Arab states but at the oil companies. And nothing the companies could say or do appeased them. The winter of 1973–74 was long and cold and tempers were short and hot. To further aggravate the situation, the companies began announcing record profits. The companies explained that 1973 profits looked so fat because 1972 profits were exceptionally lean, which was true. It was also true, percentages and yearly comparisons aside, that Exxon's 1973 profits amounted to $2,500 million, the highest of any corporation in history, and the other giants were not far behind.

The companies insisted they needed such profits to hunt for new oil, which was true. Mobil promptly created doubts about the validity of the argument by paying $500 million for Montgomery-Ward, and Gulf, already in trouble because of illegal campaign contributions to Richard Nixon, tried but failed to buy the Ringling Brothers, Barnum & Bailey Circus. Even *Fortune* Magazine chided Mobil.

But the public's indignation faded with the end of the embargo in March 1974. The shortage was over. Gasoline cost more, to be sure, but there was plenty of it. The companies went through several Congressional wringers but came out alive and kicking. The testimony appeared to be of interest only to politicians and scholars.

With all of the hullabaloo over, with a new oil surplus building, the demise of OPEC once again was widely predicted. OPEC simply cut back on its production to hold up prices and handed the majors the job of rationing supplies to the consumer nations. There were even a few sabers rattled at the Arabs, but OPEC just kept on pumping along.

There was a brief but savage gas shortage accompanying the oil shortage—also never explained to the public's satisfaction. Utility bills skyrocketed in tandem with gasoline prices. This also brought forth cries of outrage from consumers and left a residue of resentment against utility companies which may never disappear.

If nothing else, the embargo and shortage had made prophets of the independents. They had argued for decades that excessive imports dried up the risk capital necessary to drilling in the vast unexplored areas of the U.S. They had argued just as stoutly that the Federal Power Commission's control over prices of gas at the wellhead had stifled exploration for gas.

The government, in its wisdom, decreed that newly-discovered oil should bring a price competitive with OPEC's price, while "old oil" should bring an average of $5.03 per barrel. In 1975, for example, Upper Tier (new) oil brought $12.08 per barrel, Lower Tier (old) oil brought $5.03 per barrel, Middle East crude at U.S. ports brought $12.30 per barrel, Venezuelan crude brought $11.65, Canadian crude brought $12.72, and Indonesian crude, low in sulphur content, brought $13.79.

A surprising number of companies apparently could not differentiate between old oil and new, and a surprising number pleaded *nolo contendere* or agreed to settlements when the government pointed out their absent-mindedness. No one was ever accused of selling new oil as old oil.

The two-tier arrangement did not elate the independents. They insisted that all domestic crude should be priced as one, and it should be competitive with OPEC's prices. And, as always, they wanted gas freed from federal control.

Nixon, in disgrace, had handed the Presidency to Gerald Ford, his vice-president and long-time champion of the free market. Independents' hopes soared. But Ford, wanting to be President in his own right and facing a strong challenge for the Republican nomination by Ronald Reagan, signed a compromise "energy bill" that provided for continuation of oil and gas controls until 1979. It was not unusual in U.S. politics—or any politics—for expediency to rise above principle.

Republicans Nixon and Ford had failed them. Ahead was a Democrat, Jimmy Carter. It was no time to be faint of heart; the independents wiped their bloody noses and carried on the fight.

Perhaps the independents should have bowed slightly in the direction of Mecca. OPEC's action had made it possible for them to sell all the oil they could produce, and at higher prices. But an onshore well that cost $75,813 to drill in 1970 cost $150,201 in 1975, and an offshore well that cost $565,700 in 1970 cost $1,142,215 in 1975.

It appeared that if imports no longer were a deterrent to domestic exploring, inflation was.

Pete Cawthon, Jr. was born August 26, 1921 in Mexia, Texas where nine months earlier Colonel A. E. Humphreys had brought in the No. 1 Rogers from 3,060 feet to transform the drowsing community into a roaring boomtown of 40,000—a boomtown where, it was said, there was more oil than water.

Young Cawthon was not particularly impressed with the field. Later he would become a distinguished petroleum engineer, but in his youth at Mexia, Sherman and later Lubbock, his chief concerns were scholarship and football. His hero was not an oilman but "Coach Pete," his father, the prototype of the disciplinarian as football mentor. Long before Texas Tech was admitted into the Southwest Conference, Coach Cawthon's rugged Matadors regularly thrashed conference teams and independents rash enough to schedule them. His teams were tough physically and mentally. He disdained other sports as sissified, forbidding his players to compete even in basketball (which he called "thump thump ball"). When his team lost 13-6 to St. Mary's Galloping Gaels in the 1939 Cotton Bowl Classic, Coach Cawthon disappeared into the mountains of Arkansas and didn't come out until his wounds had healed.

He was back in Lubbock in time, however, to see young Pete quarterback the Lubbock High School team to a state championship. Father and son agreed that for young Pete to play at Texas Tech he would have to be twice as good as his competition. So Pete took off for the University of Oklahoma where he had been offered a football scholarship. For the first time in his life he learned what it was like to play second-string. He was too slight and slow of foot, the OU coaches said, to be a starter, so he put in his time as a reserve, ready when called on, as he was.

But Cawthon was a student who played football in order to get an education. He took up engineering because it was challenging, and petroleum engineering because it was the most exciting branch of the profession in the area where he had grown to manhood. During the summer months he worked as a roustabout and roughneck for Noble Drilling Company and Humble from the swamps of South Louisiana to the hard rock of the Permian Basin.

His education was interrupted by World War II; he entered the Army as a private and came home from Europe as a first lieutenant. Back at school, he obtained his B.S. and M.S. degrees in petroleum engineering. And he married Charlsie McLaughlin of Ardmore, Oklahoma, whom he had met at OU—and he found another hero.

While working at a summer camp "Coach Pete" was running in Virginia, Cawthon got caught up in the history of the Confederacy—and another Civil War buff was born. He became a collector of Confederate relics, and General Robert E. Lee lined up beside "Coach Pete" in the ranks of Cawthon's heroes.

Cawthon went to work for Phillips Petroleum Company as petroleum engineer in 1948. He was there less than two years when he caught the eye of George Nye and A. G. Gueymard at First City National Bank in Houston. They were looking for a younger man with a modern education in the new technologies. Petroleum banking was expanding as more and more banks became aware that loans to independents were safe loans as long as banks had professional engineers to advise them.

Cawthon worked his way up through the ranks to become Senior Vice-President, Petroleum and Minerals, First City Bancorporation of Texas, a position he reached soon after he became president of the Petroleum Club. Handsome, looking much younger than his 54 years, he brought a bright enthusiasm to the Club presidency, and a capacity for team leadership. The old quarterback made himself right at home. He was fortunate, an admirer said, in serving with directors who were leaders also, and he had been wise in his selection of committee chairmen. "If Pete doesn't know all there is to know about something, he can damned sure find somebody who does."

One of the things they did was toss out the old pension plan and produce a comprehensive, enlightened retirement program for employees that virtually halted the employee turnover problem that had plagued the Club for decades. Combined with the employee insurance program and constant attention to salary levels, the retirement program proved to be the catalyst.

Directors Lee Hill and Alf Roark spent countless hours in consultation with House Committee Chairman W. J. Mechura in devising the retirement program. Attorney Roark was a questioner: Is this right? Why this? Wouldn't this work better? Are we certain here? Does this outside expert know what he's talking about? Attorney Hill was an "I dotter" and a "T crosser:" Yes, but. Strengthen this paragraph. Move this, move that. Mechura was the workhorse. We'll have it. Don't worry. And the committee has another idea about another problem.

Mechura's committee, with Willard Johnson as the Board's liaison, studied the Club's general insurance coverage, another ball tossed to the committee by Cawthon. Mechura made a neat reception, produced a new plan to protect the quarters, contents and members, and the Board approved it. But is the staff prepared for an emergency? Roark wanted to know. Yes, said Erik Worscheh. The fire marshals come by regularly and they have taught us escape techniques. All is well.

It was brought to the Board's attention that Joe Ramirez, the Club's long-time employee, had retired before the new pension plan became effective. This was remedied. Roark said that he and others felt that Ramirez' retirement from full-time employment had gone virtually unnoticed. He offered the following resolution:

"WHEREAS, Mr. Jose C. (Joe) Ramirez has been a valued employee of the Petroleum Club of Houston since February 1, 1953; and,

"WHEREAS, Mr. Ramirez, after being employed by the Club, progressed through the years to Head Waiter, Maitre d', and in 1971 was promoted to Assistant Manager, and endeared himself to the members of the Club, their families, and visitors; and,

"WHEREAS, his conduct and demeanor throughout the years have been projected in a manner that added to the prestige, dignity, and warmth of the Club; and,

"WHEREAS, Mr. Ramirez reached retirement age on the 16th day of January, 1975, and severed his permanent, full-time employment with the Club; but

"WHEREAS, Mr. Ramirez elected to continue as a part-time employee and as hospitality captain in meeting and greeting members and guests of the Club; and,

"WHEREAS, individual members, as well as members of this Board of Directors, have expressed to him informally and verbally their appreciation of his long years of service; and,

"WHEREAS, however, the contributions of Mr. Ramirez to the Petroleum Club of Houston deserve official recognition;

"NOW, THEREFORE,

"BE, AND IT IS HEREBY RESOLVED, that the officers and members of this Board of Directors, acting in behalf of the members of the Petroleum Club of Houston, express to Mr. Jose C. (Joe) Ramirez their appreciation for his long years of employment, his dedicated service, his gracious personality, and the professional manner in which he has discharged his responsibilities;

"BE IT FURTHER RESOLVED that Mr. Ramirez be cited as one who has contributed beyond his realm of service toward the enhancement of the image of this Club; and,

"BE IT FURTHER RESOLVED that Mr. Ramirez recognize that he is the recipient of the ongoing friendships of members of the club and that their best wishes follow him and members of his family throughout their lives; and.

"BE IT FURTHER RESOLVED that a copy of this Resolution be signed by the President of the Petroleum Club of Houston, certified by the Secretary, appropriately framed, and presented to Mr. Ramirez as a memento of the high regard in which he is held by members of this Club. . . ."

The resolution recorded the beginning of Ramirez' employment as February 1, 1953. That was the date when the Club took over food and service operations from the Rice Hotel. Ramirez, as a hotel employee, had worked in the Club from its opening in December 1951.

To make certain that all members would be aware of the Board's action, a copy of the resolution was mounted in the club lobby. Ramirez' cup overflowed.

After careful study, the Board made moves to maintain the Club's financial equilibrium by increasing food prices slightly in the private dining rooms, by hiking the beverage pool tab from $4.50 monthly to $7, and by recommending that annual dues be increased from $360 to $432 for resident members, from $180 to $216 for Seniors, from $90 to $108 for associate and non-resident members.

An ad hoc committee chaired by Maginnis recommended that the limit on resident memberships remain at 1,500 but that available memberships be closed off at 1,480 with 20 memberships set aside to be assigned to applicants at the discretion of the Board. The 20 memberships would be replenished by resigned shares of stock. The Board approved the recommendation, vowing to be more selective in approving new members by giving priority to those more directly connected with the oil industry.

In advance of the annual membership meeting—at which they had been planning to obtain a vote on the dues increase—the Directors instead drew up a resolution that permitted the Board, by simple resolution, to increase the dues themselves. It was Board reasoning that dues could be increased in lesser increments, to the advantage of the membership, in such fashion.

The membership agreed, voting 724 to 108 in favor of the resolution. The succeeding administration installed the dues increase. Members also voted in favor of changing eligibility for non-resident members from 30 miles to 75 miles from Houston City Hall in recognition of the city's growth.

At that annual meeting—May 25, 1976—Cawthon, Bolton, Blanchard, and Johnson stepped down from the Board. In their places the members elected R. M. Edwards, John W. Phenicie, Eugene F. Shiels and John P. Townley. Edwards had been Cawthon's Entertainment Committee chairman, Phenicie of Amoco had been Milton Gregory's Entertainment chairman, Shiels was executive vice-president of Zapata Corporation, and Townley was with Texas Commerce Bank.

At the annual organizational meeting Lee Hill, who had retired from Exxon five months earlier, was elected to succeed Cawthon as president. Alf Roark was elected first vice-president, Gayle Whiddon was named second vice-president, C. F. Bowden was chosen treasurer, and C. F. Maginnis was elected assistant secretary-treasurer.

Hill appointed Dallas W. Johnston chairman of the House Committee, Robert Stewart, Jr. chairman of the Finance Committee, G. T. Armstrong chairman of the Entertainment Committee, Terrence McGreevy chairman of the Food Committee, Fred Heyne, Jr. chairman of the Wine Committee, and H. D. Thornton chairman of the Publications Committee. Johnston was vice-president of Barber Oil Exploration, Inc., Stewart was vice-president and assistant to the chairman of Houston Natural Gas Corporation, Armstrong was general manager of Welex, McGreevy was a former director and attorney, Heyne was a past-president and charter member, and Thornton was president of Teal Petroleum Company.

Hill had been a tower of strength in the Cawthon administration. He was the third Exxon executive to be elected to the Board and the first to be elected president of the Club. He was a complex man, brilliant and aggressive yet, as Upshaw one time said, he would "melt at the sight of a beautiful rose." Indeed, he was president of the Houston Rose Society, and the rose garden at his home was a source of constant delight and pride.

Hill was born in 1910 at China Spring, near Waco, in Central Texas, one of the few areas in the vast state where the drill had never found an oil sand. China Spring was a hamlet. Waco, in contrast, was a metropolis, and it was there that Hill went to attend Baylor University. He was graduated from Baylor in 1931 with a B.B.A. degree—and he had the highest scholastic average in the School of Business graduating class. Later he attended the Graduate School of Business at the University of Texas and obtained his L.L.B. degree from the University of Texas Law School in 1939. He had been working all the time that he had been obtaining the degrees in Austin—in Austin, San Antonio, and Corpus Christi. And he had married Barbara Camille Long of Austin. He joined Humble's law department in 1941.

In working his way up in the Humble-Exxon hierarchy, Hill had been active in many industry and tax-oriented organizations as well as civic and charitable groups.

It was a credit to Hill the individual that his elevation to the presidency by the Board brought no outcry from scattered members that Humble-Exxon was trying to take over the Club. Some had feared when the Club was first built in the Humble tower that the company would overwhelm the independents. Their fears were groundless; the company had leaned over backwards to keep from creating that impression. Hill had gained the presidency on his merits, and was duly honored by the membership. That he gave the Club his best efforts, as a member, director, and president, could not be denied.

During his administration the initiation fee for resident members was increased from $1,800 to $2,400, the fee for associate members was raised from $300 to $400, the transfer fee was increased from $120 to $600, and the stock transfer fee from $300 to $900.

And it also was during Hill's administration that a major construction program was planned and initiated. There were three reasons why the program got underway: 1) The membership quota had been reached and Club facilities were not quite expansive enough to assure comfortable service and allow membership growth. 2) Private dining rooms on the 44th floor were in such demand by members that they were booked weeks in advance, and Worscheh was forced to turn away needed business—and the private rooms already were providing 50 percent of food and beverage income. 3) Exxon wanted the space on the 29th floor occupied by the Club's accounting department. The company was willing to discuss renting space on the 45th floor that formerly had been used as a viewing tower (when the building had been the city's tallest) and was now being used as a training quarters. There was enough floor space on the 45th floor to accommodate the accounting department plus the Club's general offices currently on the 44th floor.

The construction program did not spring full-grown from the Board's brow. The Pete Cawthon administration had made a study of membership growth and space requirements, and Exxon had let it be known at that time that it wanted the 29th floor space and would dicker over the 45th floor. The Hill Administration first decided to increase the size of the Wildcatters Grill. George Pierce, the architect who had created the Club as it now stood, was engaged to make a feasibility study and the Board voted to increase the provision for the Replacement and Improvement Fund from $100,000 annually to $144,000. And Hill appointed an Ad Hoc Construction Committee composed of W. Henson Moore (chairman), Eugene Shiels, Dallas Johnston, James Teague, and Kenneth Montague.

The committee then was authorized to make a larger study of the 43rd, 44th, and 45th floors to determine just how much space could be made available to the Club. This authorization was granted in January 1977. At the April 12, 1977 Board meeting, Henson Moore reported that his Ad Hoc Committee had told Pierce to go ahead with his study of changes for the 43rd floor, and he asked for more money for Pierce to study all three floors. And he said a subcommittee had been appointed to work with Exxon on space requirements—and to discuss the possibility of an extension of the Club's base lease.

There the matter rested briefly because Hill's tenure as president terminated at the May 24, 1977 annual membership meeting. Hill, Roark, and Whiddon stepped down from the Board and were replaced by Andre Crispin, G. Will Frank and Charles E. Shaver. Crispin was president of the Crispin Company, Frank by now was a Houston Oil International executive, and Shaver was associate general counsel for Exxon.

The meeting was not harmonious. Enlarging the Wildcatters Grill meant that the bar and card room would cease to exist. There were cocktail lounges where members could gather for a drink or two, but the bar and card room in the Grill had become honored by time.

Aubrey Stautberg, past-president and long-time member, said he was opposed to doing away with the bar and card room. Hill asked that a motion not be made but that comments be noted. After other comments on other subjects by several members, Stautberg regained the floor.

He said that those present wanted to make it clear to the Board—and future Boards—that the members wanted to retain a Men's Bar in the Club. And he made a motion to that effect. It was seconded by Michael Kelly, former director and moose hunter of some renown. The motion carried. Space would have to be found.

There was no question that the Grill enlargement was needed. Business was good—too good at times—and attendance at special functions was impressive. There were no complaints about the food and service. Worscheh seemed to be introducing some new delight for the palate almost every week. Terrence McGreevy, the Food Committee chairman, noting that menu prices had been raised again because of rising food costs, was high in his praise of the General Manager for taking the sting out of the increase with his innovations. "One of the most outstanding was the new menus developed by Mr. Worscheh and Chef Lehmann which offer an outstanding variety of entrees as the menus are rotated daily in three-week cycles." McGreevy reported. "The club received an award from the Club Managers Association of America for these unique menus. . . ."

G. T. Armstrong, Entertainment Committee chairman, also had words of praise for Worscheh, pointing out that the special-event dinners were increased to 12 during the year, and every one was a sell-out. "The excellent food served by the Club staff is obviously responsible for this capacity attendance," Armstrong reported. The only "damp spot" on that entertainment calendar was caused by the cold rain that fell on the annual golf tournament.

One report to the membership during the Hill regime would have an impact on the next. It was made by Fred Heyne, Jr. as chairman of the Wine Committee. It began:

> "This has been a significant year in matters relating to wine at the Petroleum Club. I am not being effusive because the continuously mounting interest and consumption of wine is irrefutable evidence for such a statement.

"The dollar volume of wine sales for the recently ended three-year fiscal period increased 33⅓ percent and is generating a noticeable profit to the Club. This may seem astonishing to some but it comes as no surprise to those who have worked diligently on this Committee for some time and are responsible for the magnificent collection of wines we have to offer the membership at prices that are oftentimes one-half to as low as one-fourth current replacement prices.

"One of our prime objectives has always been to lend aid to the destruction of the old concept held by many that wine is for the Europeans. That, in fact, is not compatible with the American way of life. This has been relatively easy as the capacity acceptance of the several Seminars held throughout the year together with dollar volume of sales clearly indicates. We have now included a three-course meal in the format of the Seminars. This expands and complements the educational process by actually comparing several wines to each course served. It places each wine in a proper frame of reference and at the same time adds zest to the evening's entertainment. The comments regarding the addition of food by those in attendance have been consistently enthusiastic. They all encourage the continuation of this plan of presentation.

"I would like not only to commend each member of the Committee for his service but also to express my appreciation for their sincere and extraordinary dedication toward providing our Club with a well selected variety of wines with which to satisfy the tastes of all members at prices that generally are unobtainable at other places. . . ."

Heyne had spelled it out; wine was a money-maker. More and more members were learning to enjoy it at affordable prices. Wine tasting had been considered a frivolous pastime when it had been initiated at the Club in the mid-1960s with cheese and crackers to cleanse palates. As the seminars had flourished, so had sales of wine. Now, Heyne said, wine sales were contributing substantially to Club income. (Indeed, Worscheh, as knowledgeable in wines as he was in food and its preparation, had bought wines in volume for $90 a case that appreciated in value to $4,000 a case in a matter of years. In a way it was like buying CD's at an incredibly high rate of interest.)

Heyne's report would be remembered.

W. Henson Moore was elected to succeed Hill as president of the Club. A. L. Ballard was elected first vice-president; C. F. (Doc) Maginnis, second vice-president; Eugene Shiels, secretary; John Phenicie treasurer; and G. Will Frank, assistant secretary-treasurer.

Moore, as we know, had been chairman of the Ad Hoc Construction Committee in the previous administration. As president of the Offshore Company he was the first representative of the vital offshore industry to take the helm of the Club. He was aggressive, hot-tempered, abrupt at times, but widely regarded as an outstanding executive. Gueymard, who had known him since the 1930s, described him as one of the most competent businessmen he had ever known.

Moore was born in Beaumont, Texas in 1913. He attended Lamar Junior College there and the East Texas College of Law. Ahead was a 6-week cram course at the University of Texas, plus two more weeks for a bar examination—all at a cost of $800, which he didn't have. It was 1934. He landed a job as a roughneck for the Yount-Lee Oil Company at the High Island field.

In 1935 Yount-Lee sold out to Stanolind. Moore was drilling one day, feeding pipe into the hole, when the pipe grabbed. He strained his back. It wasn't a serious injury, so Stanolind put him to work in an office while he recovered. Moore recovered quickly, but his superiors had liked what he had done in the office, so they kept him there. He did it all—warehouseman, materials man, assistant to the superintendent, field clerk, district clerk, division clerk—learning all the while and developing leadership qualities.

In 1950 he went to work for Southern Production Company which, after some convolutions familiar to the oil industry, became the Offshore Company in 1954. Moore was deeply involved in the transformation. He had been head of the drilling department for both land and barge rigs, and he was active in the transition to deepwater rigs.

Offshore was a pioneer in overwater drilling. The company set record after record, registered patent after patent, under Moore's direction. It became a giant with rigs and offices in operation around the world.

Moore was a doer, and so were his fellow officers and directors and the committee chairmen selected to aid the administration.

Herbert Thornton remained as chairman of the Publications Committee. Clark Edgecomb of Schlumberger was House Committee chairman; R. A. Baile, a Berry Industries Corporation executive, was Finance Committee chairman; William V. Grisham, an Amoco executive, was Entertainment Committee chairman; Dr. Marshall Henry was Wine Committee chairman, and John B. Brock III of Brock Petroleum Corporation was Food Committee chairman.

Moore named Maginnis, Shiels, Townley, Teague, and Montague to the Ad Hoc Planning Committee with Maginnis as chairman. Maginnis also was liaison with Edgecomb's House Committee.

Moore also asked the Board to approve a new group to be named the Audit and Legal Review Committee. It was formed with Charles Shaver as chairman. Members were Maginnis; A. L. Ballard; John Phenicie; John Townley; John Cabaniss, partner in Andrews and Kurth law firm; and Grant A. Fuller, a partner with Price Waterhouse. The committee's mission was to provide the Board and the staff with review and assistance with audit and legal problems—and it was kept busy in dealings with the Internal Revenue Service and rewriting the bylaws to provide for compliance with the intent of both the Department of Labor and the Treasury Department regarding non-discrimination. (Members were no longer men but persons with the bylaws changes.)

It was a busy and productive year for all committee members, but Maginnis—as chairman of the Ad Hoc Planning Committee and liaison with Edgecomb's House Committee—seemed to work at a dogtrot as physical changes in the Club were being shaped and discussions with Exxon progressed. Had Maginnis relaxed, there is little doubt that Moore would have prodded him.

It was Eugene Shiels, Board secretary and member of the Ad Hoc Committee, who came up with a new location for the bar and card room, to the delight of those members who insisted that those oases not be blown away. Walking along the corridor from the Club entry to the Wildcatters Grill with Worscheh one day, Shiels halted at the library, which sat back on either side of the corridor. At the west end of the north indention was a storage room and a small office that was not being used.

"That's it, Erik," said Shiels. "We tear out the storage room and office for the bar, and the rest of the space can be the card room."

Worscheh was in immediate agreement because it was obvious that the bar would be in a position to also provide service to the cocktail lounge and Discovery Room, which were on the north side of the Club.

Designed by Architect Pierce, the comfortable area would be named the Presidents Bar, and portraits of past presidents would hang on the wall. And, at Shiels' suggestion, the room would be discreetly screened off from the corridor.

The minutes for the Board meeting of May 9, 1978, summed up succinctly almost a full year of work by the House Committee and the Ad Hoc Planning Committee. The minutes said:

> "Mr. Maginnis, in reporting for the House Committee, advised the Board that the lease extension with Exxon covering the Club's quarters had been approved by both parties and executed, the lease providing for an additional 15 years of occupancy of the Club's quarters commencing April 1, 1978, and terminating March 31, 1993, with the Club having an option for a further 10-year extension beyond that date.

> "The question of escalation in rental rates was fully covered in the new lease with annual rentals commencing at approximately $190,000 per year and increasing to a level of about $290,000 per year for the last 10 years. In addition, the Club has an obligation to Exxon for a pro rata share of any future cost of maintenance and repair to the heating and air conditioning systems, such share being 3.9% of the total cost, as the Club's net rentable area represents 3.9% of the total net rentable area in the building. Additionally, the Club in the future will be liable for the repair of all plumbing and any changes occasioned by leaks in the plumbing except that plumbing that relates to toilets and washrooms, which remains Exxon's responsibility.

> "Mr. Maginnis further advised that in connection with the move to the 45th floor of the Club's offices, Exxon had agreed to give the Club a $10,000 credit to apply against the costs incurred by the Club in rearranging partitions, electrical outlets and the like on the 45th floor. He further advised that the House Committee would meet on May 10, 1978 (the next day), to consider a budget

for the move of the Club's offices as well as to consider a proposal to redecorate the ladies' lounges on the 43rd and 44th floors.

"Mr. Maginnis also reported for the Ad Hoc Planning Committee that the construction documents had been approved, and that most of the furniture had been ordered after approval of the overall plans by a meeting of the Ad Hoc Planning Committee. The only remaining details had to do with the revision occasioned by the addition of the Wine Room. Final plan revisions in this connection are scheduled to be submitted by the architect to the contractor within the next few days, following which a cost estimate for these changes should be forthcoming.

"Mr. Maginnis advised the Board that both the Ad Hoc Planning Committee and the Wine Committee members who consulted with the House Committee were highly pleased with the concept of the Wine Room as developed by the architect, adding that it would provide additional private dining space for as many as 24 guests in surroundings that were appropriate and yet unique in that they provided on the 44th floor badly needed storage space for wine, the capacity in the Wine Room being 256 cases. . . ."

If this section of the Club minutes seemed unduly interested in the Wine Room, it must be remembered that Secretary Shiels was both a member of the Ad Hoc Planning Committee and the Wine Committee. There had been no provision for a Wine Room in the original plans for the 44th floor. The Club's considerable stock of fine wines was kept in a basement of the building, far removed from the Club, and only small amounts could be held on the 44th floor when a wine tasting or seminar was in progress. The "winos" envisioned a room that could be both a cellar—with a constant temperature—and a beautiful room for tastings and seminars . . . and a room that would be available for other parties and functions when the "winos" weren't using it.

They were aware that Fred Heyne's report on the value of wine sales to the Club had not gone unnoticed by the Board and by the membership. They also were aware, however, that President Moore in the past had made some jocular remarks about the Wine Committee which could have been considered uncomplimentary. Moore was invited to a dinner for the Wine Committee and the Food Committee. To coin a cliché, Moore was royally wined and dined.

He stood up and addressed his hosts. "I never thought I'd say this, but I really enjoyed the wine," Moore said, "and I want to say that I think the Wine Committee is doing a good job."

Thereupon, the "winos" came up with a change in the plans for the 44th floor that provided for their dream room, Vintage I. Some of them produced a design for the room that bore a remarkable resemblance to the Czechoslovakian party-cellars of Worscheh's youth. The idea was mercifully dispatched, and Pierce came up with a design as handsome as the Club itself, and everyone was satisfied. It was this design that was recorded in Shiels' club minutes.

Moore's administration had met its goals of going forward with the modifications to the Grill, relocation of the bar and card room, conversion of vacated space on 44 into private dining rooms, and obtaining an extension of the lease to run until the year 2003 at a reasonable price. The lease extension was a necessity in view of the capital investments required for the modifications, and especially since the old lease would have expired in 1983. (It would have cost a mind-boggling fortune to duplicate the Club in any of the newer towers in the city. And despite the additions to the city's skyline, the view from the Club in the Exxon tower was still superb. Indeed, the latest entries in Houston's skyline derby enhanced the city's beauty to those who watched from the Club's great windows as the buildings sprouted and flourished.)

H. D. Thornton, chairman of the Publications Committee, could report that a new membership directory had been distributed during the year, and he could boast that the "Sunburst" had won first place in state-wide competition for newsletters. It was a coveted award.

And John Brock III, chairman of the Food Committe, could report that Worscheh had received the "First Place Award for City Club" outstanding luncheon and dinner menus in the nation. It was the second of many awards Worscheh and his kitchen staff would win in years to come. Brock's committee made suggestions freely as to foods to be added to the menu for the buffet in the Wildcatters Grill, and they were enthusiastically accepted by the membership.

Edgecomb's House Committee also saved the Club about $100,000 in replacing some 2,700 square feet of carpeting. The original carpet was wool and had been replaced in the past by wool carpet at about $45 a square yard. The carpet had to undergo frequent shampooing because of almost daily staining by food and drink and damage from cigarette burns and ashes. Edgecomb called on Winston McKenzie, a committee member, for help. McKenzie was a wholesale furniture distributor and gallery operator. He arranged a meeting with representatives of carpet distributors. They showed their wares to the Committee. The Committee decided on a non-wool carpet, almost identical to the original carpet in appearance and, with Board approval, paid $30,000 for 2,700 square feet of it. It wore as well as the wool and cleaned as well—and cost about a quarter as much.

Moore would leave the task of overseeing the club renovation to the succeeding administration. He and Maginnis and Ballard and Bowden stepped down from the Board at the annual meeting of May 23, 1978, and were replaced by G. Turner Armstrong, John Brock III, Roy M. Huffington, and Clyde E. Willbern. Armstrong was the Welex executive, Brock had been chairman of the Food Committee in Moore's administration, Huffington was president of Roy M. Huffington, Inc., and Willbern was a Getty Oil executive.

At the organizational meeting the next day John Phenicie was elected president of the Club. Shiels was elected first vice-president; Edwards, second vice-president, Willbern, secretary; Townley, treasurer; and Shaver, assistant secretary-treasurer.

There were exciting days ahead for the Club. There was excitement in the oil patch, too, and Phenicie's outfit, Amoco, was producing much of it. And there was excitement of another sort in Washington where President Jimmy Carter was struggling to create an energy program that would be all things to all men.

CHAPTER EIGHTEEN

ON JUNE 19, 1976, MIKE HALBOUTY, the Club's sixth president, spudded in the James I in Payette Basin, Idaho. For years
Halbouty had insisted that large oil reserves existed in untried areas of the western states, and he talked interminably about great
pools to be found in the Utah-Idaho-Wyoming "thrust belt," a region of twisted formations paralleling the eastern edge of the
Rocky Mountains. Now he was putting his money where his
mouth was.

The James I was a dry hole at 14,006 feet. The bit had encoun-
tered six oil and gas shows on the way down, but none was significant.

The very next day after Halbouty capped and abandoned his
duster, Amoco (Phenicie's outfit) brought in the discovery well of
the Ryckman field in the southwest corner of Wyoming—across
the state of Idaho from Halbouty's well—and opened up the
"thrust belt." Halbouty's well was not in the thrust belt, but it was
in rank wildcat territory, as were his dusters in Fallon Basin, Nevada, and Harney Basin, Oregon.

Amoco's discovery set off a boom in the region. Ryckman was
the first of many "thrust belt" fields—Pineview, Yellow Creek,
Whitney Canyon, South Chalk Creek, Dry Piney, Tip Top,
Hogsback. It was just like the good old days. The "thrust belt"

became, in the words of the oil trade magazines, "the hottest oil play in the lower forty-eight." With "new oil" bringing about $11.50 a barrel, the boom was hot enough to make some enterprising companies ignore the inflation that now had driven up the cost of drilling a well to about $200,000.

In Washington the independents, recognizing that there was no hope for decontrol with Carter as President, asked that the price for oil from new wells be regulated to meet but not exceed "the United States' delivered price of foreign imported oil." This was a back door way of saying that they anticipated a price hike by OPEC, and they wanted a price hike at the same time.

And in December 1978 OPEC hiked the price. It was unexpectedly steep. Now the independents pleaded for an energy policy that would immediately allow world oil prices for every barrel produced from new domestic wells completed after January 1, 1979. They didn't get it.

What they got—what the industry and the nation got—was a plan from President Carter that called for phased decontrol of oil prices beginning June 1, 1979, and ending with the expiration of price control authority on October 1, 1981. But Carter proposed to a willing Congress a "windfall profits tax" to prevent "unearned, excessive profits which the oil companies would receive as a result of decontrol and possible future OPEC price increases."

The tax money would be earmarked to help low-income families who could least afford energy price increases, to increase funding for mass transit, and to finance a program of new energy initiatives and investments to develop alternatives to imported oil, the President said.

All segments of the industry cried out against the tax. It was an excise tax having nothing to do with profits, they argued, and it simply would take capital from the producer and give it to the government. The industry had to have that capital to find the production the country desperately needed. Keep tabs on us, oilmen said. If we are wasteful with the capital, if we spend it outside the industry, expose and punish us.

And while the Congress wrangled over the tax bill, Iran was shaken by revolution. The Shah fled the country. Iranian oil ceased to flow to the U.S. and elsewhere. Imported oil prices rose.

And the Congress, in its wisdom, passed the "windfall profits tax" bill.

But with all the political haggling, revolution, taxes, and other economic disruptions, rising crude prices, which had helped set off the regional boom in the "thrust belt," were heating up the entire oil patch, and more and more it looked like the industry was in for a rip-snorting good time.

More, the Congress had passed the Natural Gas Act of 1978 which totally deregulated "deep gas"—that found below 15,000 feet. Price restrictions were modified greatly on gas from "tight sands" where expensive fracturing was required, and "old" gas was allowed to increase in price in accordance with the rate of inflation.

All in all, it was just the proper time for a Petroleum Club administration to be concerned with remodeling the premises. But, as the old saying goes, that's Oil Biz.

John Phenicie was born right outside of Houston in Sealy, Texas in 1924. He grew up in Sealy, graduating from Sealy High in 1941. He attended Texas A&M and the University of Texas before being graduated from Rice Institute in 1947 with a B.S. degree in mechanical engineering. And he had served in the Navy during the war, attaining the rank of Lieutenant (jg).

Phenicie joined Amoco right after graduation as a Junior Petroleum Engineer in the West Edmond Area of Oklahoma, but soon returned to the Hastings District southeast of Houston. Later he became Field Engineer in the Hastings Area; Petroleum Engineer (SG) in the Houston Division; Reservoir Engineering Supervisor in the company's Tulsa general office; Assistant Division Engineer in the Houston Division; Division Engineer and Assistant Division Production Superintendent in the Casper, Wyoming Division; Assistant Division Production Superintendent in the Houston Division, and Production Manager in the Tulsa general office.

In 1963 Phenicie was named Amoco's Vice-President of Production where he served until he was appointed Vice President and Houston Division Manager in December 1972. He later was named Vice-President, Houston Region, the position he held when he was elected president of the Petroleum Club. He had joined the Club in 1973.

The Phenicie administration retained Herbert Thornton as chairman of the Publications Committee, marking his third successive term on the job, and also retained William Grisham as chairman of the Entertainment Committee.

J. W. O'Keefe, treasurer for Texas Gas Exploration Corporation, was named chairman of the Finance Committee; Thomas D. Gholson, president of Tubular Sales, Inc., was named chairman of the Food Committee; past-president Milton Gregory was named chairman of the Wine Committee, and E. C. (Ned) Broun, Jr. was named chairman of the House Committee. Broun, president of Petroleum Service Group, Dresser Industries, had served on the previous House Committee, and was conversant with the work to be completed. Charles Shaver continued as chairman of the Audit and Legal Review Committee.

Eugene Shiels was made chairman of the Ad Hoc Planning Committee. He was joined by fellow board member Roy Huffington and House Committee Chairman Broun. Past-presidents James Teague and Kenneth Montague continued in their advisory capacities.

Remodeling the Club was serious business, but it had its humorous moments also. On Pierce's drawings of the 44th floor a new dining area was labeled "Oklahoma Room." It had been a peculiarity of the drilling business in an earlier day that it had been practically impossible for a Texas contractor to get a drilling job in Oklahoma while Oklahoma contractors operated quite freely south of the Red River. "Why in the hell should one of our rooms be called Oklahoma?" demanded a few with long memories. "You'll never find a 'Texas Room' in an Oklahoma club."

It was pointed out that the Club already had a "Louisiana Room," and no one complained about that. Why shouldn't the Club have two rooms named for neighboring oil states?

The good-natured grousing subsided, but a warmer exchange of opinions developed when the color renderings showed that the room was to be walled and carpeted in the burnt orange coloration predominant in the Club. This time it was University of Oklahoma graduates and others who complained. They argued that the room should be decorated in white and red, as in the "Big Red" of the Oklahoma Sooners, not the burnt orange of the University of Texas. But as before, when the Board held firm, the grumbling died away. Name and decor went unchanged.

For several years Exxon had complained that water leaked from the Club's kitchen area down to Exxon offices. Various measures had been taken to halt the leaking, but now the Board voted to spend $16,000 for an epoxy job to waterproof the kitchen area.

It was decided also to replace the kitchen grease traps which had been eroded by cleaning caustics. In doing so it was found that some of the cast-iron piping adjacent to the traps was eroded also. It was paper-thin in places. It was a week-end job, and workmen were laboring beneath the kitchen in the Exxon offices. On Saturday afternoon Shiels, the Ad Hoc Planning Committee chairman, was called at home. He was told about the drain pipes. There was no time to waste, if the job was to be completed before Monday morning. Shiels told the workmen to replace the piping that had been exposed.

The Board authorized payment for the work, but Shiels told his fellow Directors that he feared that much more of the 15-year-old piping was in bad shape.

Meanwhile, Broun's House Committee recommended and the Board approved a dental insurance plan for employees and hospitalization benefits for retired employees and their dependents. Broun reported: "I feel our employees not only appreciate these benefits, but will reward the members with more enthusiastic service than in the past. . . ."

The Board also heard from Logan Bagby, Jr. Phenicie read a letter from Bagby suggesting that "distinguished lecturers" speak to Club members from time to time. "These lecturers would have to be outstanding men," Bagby wrote, "and the Committee in charge would need to be knowledgeable and influential. . . ." The suggestion had merit, but the Board—after consultation with the Entertainment Committee—voted against it for reasons of time and space. Bagby's suggestion called to mind the brilliant talks arranged by the defunct Library Committee in the old Club.

Some members had been urging that the Club apply for an Alcoholic Beverage License from the state so the Club could buy wine and liquor direct from wholesalers instead of going through retailers, as was required under the Beverage Pool System.

The proposal provoked a debate that continued for several years. At one point legal counsel was called in to advise on the matter. It was finally laid to rest when an administration decided that a change from the Beverage Pool System would be more trouble than it was worth.

The Board noted that Worscheh had again received the first place award presented at the National Convention of the Club Managers Association of America, for outstanding luncheon and dinner menus. Worscheh was commended for his achievement.

Worscheh brought before the Board Frances Harris, Membership Secretary, who was taking early retirement after 21 years of service. On behalf of the Board and the membership, Phenicie thanked her for her loyalty and cheerful willingness to do whatever she could to serve the Club's best interests. Other Directors expressed their thanks for a job well donw, and wished her continued happiness and good fortune. Her position was taken by Judith Boyce, who was introduced to the Board by Worscheh.

At the Board meeting of February 13, 1979, Will Frank, liaison with the secret Membership Committee, told his fellow Directors that he was concerned over the trend toward company-owned memberships in relation to individually-owned memberships. The trend, said Frank, was continuing at an increasing rate.

Frank had struck a nerve. The Directors had become aware that despite their best efforts to screen applicants, it had become almost impossible to control who became a member and who didn't. When a company wanted to transfer a membership from one employee to another, for example, it was difficult to question the company's judgment.

Another thing: The industry was in a state of flux, with executives moving from one company to another. If a good Club member with a company-owned membership left the company, he couldn't remain a Club member; he had to get on the waiting list to become a member again.

The Board resolved that

> "It is the sense of the Board of Directors that the Membership Committee make a diligent effort to restrain the apparently increasing percentage of company-owned memberships versus individually-owned memberships so as to ensure against dominance of the Club by corporate oriented memberships as opposed to individual or non-corporate memberships. The Board requests that the Membership Committee use its best efforts to restrain corporate-owned memberships from exceeding approximately 75% of the total Club memberships."

The irony of the situation was not lost on old-timers. For the first half of the Club's life they had tried in vain to get big companies to buy company-owned memberships for their executives. Now the Board felt compelled to limit the practice.

Soon after, John Townley, liaison to J. W. O'Keefe's Finance Committee, presented a resolution increasing the transfer fees on company-owned memberships. If an individual assigned the membership was being transferred out of the city, or was leaving the company for any reason, then the transfer fee would be raised from $600 to $1,000. If the membership was transferred to another nominee with both persons remaining with the Houston organization, the fee was increased from $900 to $1,300. The Board approved the resolution.

The Board also approved resolutions increasing resident membership fees from $2,400 to $3,000 and associate membership fees to a top of $1,500. Resident dues per month were increased from $40 to $48, associate dues from $10 to $24, and non-resident dues from $10 to $12.

In addition, the Board resolved that Senior Members who had reached 75 years of age, and who had been Seniors for not less than five years, should have their dues reduced to $3 per quarter and their Beverage Pool Contribution reduced to $6 per quarter. It was agreed that Seniors who had resigned within the past two years be allowed the privilege of being reinstated without paying dues for fees for the period of inactivity.

At the Board meeting of May 8, 1979, Shiels, as liaison with the House Committee, reported that Broun and his members were still concerned about the piping underlaying the kitchen area. More work had been done, and more studies made, but Shiels felt that it might be necessary during the next year to replace all or much of the piping. It would be an expensive undertaking, he implied.

At the annual membership meeting of May 22, 1979, the membership praised Phenicie for his work as president, and later he was presented an over-under Belgian Browning shotgun in appreciation of his service. The Board also was praised.

Phenicie, Edwards, Shiels, and Townley stepped down as Directors. The membership elected to the vacated seats John C. Jacobs, Jr., Carey B. O'Connor, Robert Stewart, Jr., and Elbert Watson. Jacobs was vice-chairman of Texas Eastern Corporation, O'Connor was a real estate broker, Stewart was vice-chairman of the Bank of the Southwest, and Watson was executive vice-president, planning, for Houston Natural Gas Corporation.

At the annual organizational meeting Will Frank was elected president; Andre Crispin, first vice-president; Charles Shaver, second vice-president; John Jacobs, Jr., secretary; Robert Stewart, treasurer; and G. Turner Armstrong, assistant secretary-treasurer.

Every day Will Frank rode his bicycle 15 miles from his home in Ottawa, Kansas to roustabout in an oilfield during his high school summers, and pedaled 15 hard miles back. The Brundred Oil Company was waterflooding stringer fields that were producing from 500 feet to 1,000 feet, and some of the wells had wooden sucker rods that were pulled by hand. By his senior year, Frank was roughnecking on a cable tool rig.

He entered Ottawa University, but World War II came along. Frank tried to get in the Navy Air Force, but a football knee kept him out. So he entered the Navy Reserve V-12 program at the University of Kansas. After almost two years, he was sent to Columbia University to attend midshipman's school. Then he went to sea, serving in destroyer escorts and submarines. He came out of the war a lieutenant. Back to the University of Kansas he went, majoring in geology and petroleum engineering. He received his degree in 1947.

He got a job with Phillips Petroleum in Eureka, Kansas, moved to Odessa, Texas, to Jackson, Mississippi (as district engineer), and then to Alvin, Texas and the Chocolate Bayou field, a producer of gas condensate.

Frank had never seen such huge Christmas trees. The man showing him around said that pressure on the low side of the field stood at 4,500 pounds, pressure on the high side at 9,000 pounds. "This is the low side," the man said. Frank marveled. "You'll be working on the high side," the man said.

In 1950 Frank left Phillips and went to work for Progress Petroleum as "the engineer, period." Progress belonged to the Andrew Weir Shipping Company. In the past Andrew Weir had owned companies like Texas Crusader, Louisiana Crusader, and Euro-Mex. EuroMex produced Mexican crude, shipped it to the Ruhr Valley of Germany, refined it, and traded the output for steel. It lost the production when Mexico nationalized the oil industry, and lost the refinery during World War II. Andrew Weir sold off the Crusader companies and retained Progress Petroleum, which had oil and gas production in West Texas and on the Gulf Coast. Frank was Progress Petroleum's general superintendent when he left in 1956 to work for Austral Oil.

He was Austral Oil's executive vice-president, production and exploration, when the company sold out to Superior in 1978. Frank then became president of Houston Oil International, a subsidiary of Houston Oil and Minerals, with interests in the Emirates, Tunisia, the North Sea, Colombia, Gabon, and the Ivory Coast. He held that job when he became president of the Petroleum Club. (In 1981, Houston Oil and Minerals would merge with Tenneco, and Frank would become Vice-President, General Manager-International, Tenneco Oil Company.)

Frank was a pleasant but forceful man, and his administration moved boldly into some new areas of Club life. Aiding the Board were Ned Simes, Entertainment Committee chairman; Thomas Gholson, Publications Committee chairman; James H. Kerr, Jr., Wine Committee chairman; William Grisham, Finance Committee chairman; Herbert Thornton, Food Committee chairman; E.

C. (Ned) Broun, Jr., House Committee chairman; Charles Shaver, Ad Hoc Planning Committee chairman and liaison, and Clyde Willbern, Audit and Legal Review Committee chairman and liaison. Simes and Kerr were new to chairmanships. Simes was president, chairman of the board and chief executive officer of Diamond M Company, and Kerr was a partner in the law firm of Fulbright and Jaworski.

Director John Brock was named liaison with the secret Membership Committee, and it was in this committee's area of responsibility that the administration cut new ground. Frank, it will be recalled, had been the Membership Committee liaison in the Phenicie administration, and he had prompted the resolution calling for restraint in the issuance of company-owned memberships. Now Frank asked Brock to make a thorough study of the Club's entire membership program.

Brock and the Membership Committee went to work. Over a drink at the Club, Brock told Frank what the study had produced, and he offered some startling recommendations. Frank was in total agreement with the recommendations and, when the recommendations were duly presented, so were the other members of the Board.

The Board unanimously approved three recommendations:

1. *There will be no new Associate and Non-Resident memberships granted.* Pending applicants will be so informed. Any applicants wishing to re-submit for full Resident membership will be welcome and the Membership Committee will take into account the time spent pending as Associate and Non-Resident applicants.
2. *There will be no new company-owned memberships granted.* Pending applicants will be so informed.
3. Present Associate members . . . will be "grandfathered" and allowed to convert to full Resident membership at age forty. *However, they will be required to take the membership as individually-owned, that is, not company owned.*

But the Board wasn't through. Also approved was a recommendation that would require a change in the bylaws by a vote of the membership. It was in two parts:

1. *Combine the two present classes of membership, Resident and Associate, into one class, Resident, with a quota equal to the sum of the two present quotas, that is, 1,600.* We will then have a present Club membership of 1,537. There will be space allocated for the present Associate members when they reach age forty, and they will not displace a pending applicant. We will then have available 63 new memberships. We will propose that only a portion of these, say 20 or so, be recommended this year by the Membership Committee.

2. *The automatic conversion of Non-Resident members will be abolished, and the quota of Non-Resident members will be established at some lower level, that is, 400–500.* The Membership Committee will take into account a new Resident's time as a Non-Resident member in considering his Resident membership application.

The Board anticipated some argument from sections of the membership once copies of the resolution were mailed out, but none arose. However, at the Special Meeting of Resident Members on November 13, 1979 to tally the vote, Robert S. Hinds, a Vinson Supply Company executive, asked to be recognized.

Hinds was eloquent and his argument cogent. He was upset about abolition of the Associate Member category. He moved that the following resolution be presented for a vote:

"WHEREAS, there is a desire to eliminate the Associate Member Classification after all present Associate Members are assimilated into the Resident Membership and,

"WHEREAS, there are present applicants for Associate Membership who applied long before there was any known proposal to eliminate the Associate Member Classification and,

"WHEREAS, both the applicants and their sponsors acted in good faith at a time when persons qualified by existing criteria would expect acceptance as an Associate Member,

"BE IT RESOLVED, in the interests of fair play to the applicants and the Petroleum Club members who sponsored them, that all applicants for Associate Membership prior to this meeting be processed in the manner prevailing prior to this meeting, and all presently qualified Associate Member applicants be accepted prior to the final implementation of the bylaw change."

Hinds didn't get a second for his motion, so his resolution was not presented for a vote. The amendment to the bylaws passed by a vote of 904 to 15. But Hinds was not forgotten by Will Frank. Frank brought up Hinds' name at several Board meetings after the vote was cast, and both Frank and Brock met with him. Hinds had sponsored a young man and felt committed to his promise to help. Frank and Brock—and the other Directors—sympathized with Hinds, and his protege, but they agreed that what had been done was done and over, and for the best.

The Board and the Membership Committee felt that the Associate Member category had served its purpose, that its days of usefulness were over. Back in the 1950s Bob Smith, Hugh Q. Buck, and others had seen it as a way to build for the future. Now, the Board reasoned, the future had arrived. The new memberships now opened by the consolidation could be allotted to explorers and producers who had been on the Resident waiting list too long.

Even so, there would still be a waiting list, and each applicant would be more carefully screened than ever before. The Club was in the position of accepting only the best qualified for membership.

In this connection, the Membership Committee and the Board were faced by their first lady applicant, an attractive, intelligent chemical engineer. She had strong support from a Board member.

The Club had its share of dinosaurs. The oil industry, for all of its internecine warfare, was tightly knit toward the world at large, politically and ideologically. The Club never would have been mistaken as a launching pad for the Women's Liberation Movement. Ladies as welcomed guests were one thing. Ladies as members was something else, for a large segment of the membership.

Brock and the Membership Committee talked it over, then Brock talked it over with Frank. It was decided that if the lady had qualifications equal to the men on the waiting list, she could get in line.

Brock took the lady to lunch. During the interview it was revealed that the lady was married but chose to retain her maiden name. Her husband's name rang a bell with Brock; it was on the waiting list. Brock suggested that the lady discuss the matter with her husband, who also had excellent qualifications. Brock explained that the list dwindled slowly, that the Club was in a position to accept only the applicants with the best report cards. It might be better, he suggested, if only one of them applied. "It will be me," the lady said positively.

It was neither, as it turned out. Both took themselves out of contention, for reasons of their own.

While the changes in the membership structure were being made, the Finance Committee recommended increases in dues, and the Board approved them. Business was at an all-time high and Worscheh, granted the authority by the Board, increased menu prices when he thought it was prudent. But employees were given additional health insurance benefits, and inflation continued to be a specter at every Club function.

It had cost a lot of money—well more than $600,000—to remodel the three floors during the Moore and Phenicie administrations. But now it had become apparent that the 17-year-old Club was threatening to come unglued at some of its seams. The expensive dimmers in all the rooms needed replacing. Some of the operating equipment had seen its best days. New casements were needed on the 43rd floor. And worst of all, it had become evident that all of the lateral drain lines were shot because of piping leaks. Band-aids wouldn't work, House Committee chairman Broun told the Board, and just replacing the lines would cost as much as $250,000.

The Board didn't want to disturb funds invested in CD's. Dues had just been increased, and it didn't seem proper to increase them again so soon. So, on recommendation of the Finance Committee, the Board voted to slap an assessment on the membership—the first in Club history.

Assessment had amounted to a dirty word in earlier years. Boards had reasoned that members would resent an assessment when the Club always had a reserve fund. This probably was a correct estimate when the membership quota was not full, and the Club was struggling to keep the members it had. But now there was a waiting list. And despite President Carter's "windfall profits tax," the oil business was booming.

Resident members were assessed $200 each and Non-Resident members were assessed $50 each. And, as recommended by the Finance Committee, the Board also voted to increase new Resident membership fees from $3,000 to $4,000 and new Non-Resident membership fees from $300 to $400.

So bankrolled, Broun's House Committee signed the necessary contracts to get the project rolling. And at the annual membership meeting, President Frank called on Broun to report on the lateral drain system. Broun did more than that; he held up sections of corroded piping to dramatically demonstrate the problem. The Club, Broun said, was responsible for repair and replacement of the system, according to the lease with Exxon. He said he knew the assessment was a sudden blow, but he pointed out that under the lease contract the Club had a rental—and quarters—that could not be matched anywhere in the city.

Frank explained that other methods of funding had been explored, but the Board had concluded that the assessment would provide the necessary money most expeditiously. Broun's pipe-waving, his speech, and Frank's explanation carried the day; there was no apparent ill-feeling about the assessment.

A non-member could not help but be impressed by the pride members had in the Club. For the old-timers, it had never waned from opening day. Newcomers seemed to inhale it like incense once they had a Club card in their wallets. Ned Simes, Entertainment chairman, may have summed up the feeling in a report to the membership. He wrote that more than 500 had attended the Second Annual Petroleum Club Ball, and added:

> "It is estimated that we need 600 in attendance for this function to be profitable; however, there are certain activities that are associated with a club of our magnitude that must be continued in spite of 'losing money' on the function itself. . . ."

Frank, Shaver, and Crispin stepped down from the Board at the annual meeting of May 27, 1980. They were replaced by Broun, two-time House Committee Chairman; Thomas J. Feehan, president of Brown and Root, and Thomas W. Stoy, vice-president, Gulf Region, Union Oil Company of California.

At the organizational meeting, John Brock was elected president. Other officers elected were Roy Huffington, first vice-president; John C. Jacobs, Jr., second vice-president; Elbert Watson, secretary; Robert W. Stewart, Jr., treasurer; Clyde Willbern, assistant secretary-treasurer.

Director Carey O'Connor had died on the day before the annual meeting, and the new Board appointed H. Ross Bolton, owner of HRB Oil, to serve out O'Connor's term.

The adminstration selected John T. Cooney as chairman of the Finance Committee, William Grisham as chairman of the Food Committee, C. D. Roxburgh as chairman of the Entertainment Committee, Harry M. Jacobson as chairman of the Publications Committee, and Eugene Shiels as chairman of the Wine Committee. Clyde Willbern continued as chairman of the Audit and Legal Review Committee, and John C. Jacobs, Jr. was named chairman of the Ad Hoc Planning Committee. H. J. Gruy, who had worked with Broun on the previous House Committee, was now named to chair it.

Newcomers to chairmanships were Cooney, Roxburgh, Jacobson, and Gruy. Cooney was an executive with Bank of the Southwest, Roxburgh was an Exxon executive, Jacobson was assistant secretary of Brown and Root, and Gruy was chief executive officer of H. J. Gruy and Associates, Inc.

John Brock, at 48, was one of the youngest men to be elected president of the Club (Marlin Sandlin was 44). But he was seasoned, both in the oil patch and in command of troops along the Rhine River during the coldest days of the Cold War, when West Berlin was being provisioned by the American airlift and the sun rose each morning on a tense, uneasy world.

Brock was born in 1932 in Melville, St. Landry Parish, Louisiana, a thriving town where traffic from north Louisiana was funneled to south Louisiana and New Orleans over the Atchafalaya River. In that year, St. Landry Parish did not vote for Huey Long for governor. Elected anyway, Long re-routed the highway 10 miles south and Melville went into a decline. And so it goes.

Brock went to grammar school in six different Louisiana towns before his father, a civil engineer in the construction business, settled in Lafayette. Brock grew up there. He was a high school athlete, a track man, becoming the state record holder in the half-mile.

At Louisiana State University Brock ran the quarter mile and the half mile, and was on the conference championship mile relay team. On his way to a degree in petroleum engineering, he worked summers as a roughneck for Delta Drilling Company. He also was in the university's ROTC unit.

On graduation Brock went to work for Humble as a trainee engineer. He worked only three months; the government called him to active military duty. He spent a year in training and teaching, then a tough year guarding the bridges along a 100-mile stretch of the Rhine. His orders were simple: let our people fall back across the bridges, then blow them up; put pontoon bridges in the river for our people, and when our people cross them, destroy the pontoons.

Back home Humble told Brock the company wanted young engineer trainees to roughneck 18 months on an Humble rig and roustabout another 18 months in an Humble crew. Brock's roughnecking during the summers didn't count, the company said. Humble changed this policy a few months later, but by then Brock was working as a petroleum engineer for the California Company, a subsidiary of Standard of California.

He was hired in New Orleans. "You know where Lafayette is?" the man asked. "I grew up there," said Brock. "Know where Nat Mouton's Battery Shop is?" "Across from Don's Seafood Inn," said Brock. "There's a stairs behind the battery shop," the man said. "Climb 'em, and there's our office."

Brock climbed the stairs to the little office and met more questions, these from Harvey Fitzpatrick, a brilliant engineer. Fitzpatrick gave Brock a manual. "Know anything about the oil-field?" he asked. "I really don't," said Brock. "Know where Franklin is?" "Yes," Brock said. "Know where the sugar cane mill is?" Brock said he did. "Know where the bridge over Bayou Teche is?" "Yes, I do," said Brock. Fitzpatrick tossed Brock a set of keys. "There's a green Chevrolet parked downstairs. Go to Franklin. Go the sugar mill. Go to the bridge. Turn right and go about a mile. You'll find a drilling rig. They're getting ready to come out of the hole to log. A geologist is coming out there. Yawl study the log, then give me a call and tell me what you want to do."

"Is that it?" Brock asked.

"Yes. Read that manual. It covers everything, but you've got to think. Call me anytime you want to, but don't call me until you know what you want to do."

Brock was 23. He worked for Fitzpatrick and California Company for three and half years. Within six months after he started he was running six or more drilling rigs. He was a drilling engineer, then production engineer as the company sank deep wildcats in South Louisiana.

Quintana had opened a small office in Lafayette with a geologist and a production superintendent. The company was looking for a district engineer. It had some production in the area and had just discovered what the geologist thought was a good gas field— Garden City. Brock went to work for Quintana.

Garden City turned out to be a giant, with wells clocking 13,100 psi shut-in tubing pressure. It was a learning ground where new technologies were developed, and Brock had six engineers working at his direction.

Brock came to Houston in late 1963 as Quintana's chief engineer. Later he was assistant production manager. And he spent two years on a Quintana copper mining operation near Tucson, using oilfield techniques to core a deep, rich vein. Finally he was given responsibility for all of the company's oil and gas business.

A month after he became Club president in May 1980, Brock struck out on his own, forming Brock Petroleum Company. It seemed a likely time; the oil business by now was caught up in a boom almost beyond belief, and everybody's bits were on the bottom and turning to the right.

CHAPTER NINETEEN

WHAT A WONDERFUL boom it was! For majors and near-majors, for independents large and small, for oilfield manufacturers and oilfield suppliers, for wildcatters, consultants and outlanders who didn't know a derrick from a hat rack! For banks and brokerage houses with formalized funds, for widows, orphans and misers with mossy-green nest eggs! Possibly no more than 50 U.S. banks had the expertise to make oil and gas loans to independents, but now it seemed as if all 14,000 banks from coast to coast were trying to shove big bills into independents' pockets—and it is an oil patch axiom that a good independent will borrow as much money as you will lend him!

Like the material in a breeder reactor, the boom fed on itself. By mid-1980 Arabian crude was bringing $30.02 per barrel at U.S. ports. Canada and Mexico were not OPEC members, but they were not averse to OPEC prices; Canadian crude brought $30.47 per barrel and Mexican crude brought $31.80. Domestic crude prices tagged behind, but not so far that there wasn't a rush to find and produce it.

And gas! "Deep gas" was worth more than $9 per mcf to the transmission companies and other buyers who signed up for reserves on a "take or pay" basis. Other gas was bringing as much as $5 per mcf.

In addition, Ronald Reagan, running as the Republican candidate against Jimmy Carter, was strongly opposed to all controls on oil and gas and the hated "windfall profits tax." He was elected, and immediately decontrolled oil. Domestic crude prices surged upward. But with a military-civilian budget without precedent in war or peace, the "windfall profits tax" remained untouched, and total decontrol of natural gas got placed on a back burner once again.

Meanwhile, the price of regular gasoline had gone from about $0.88 per gallon to $1.22 to $1.31. Home heating oil had risen from about $0.73 per gallon to $1.01 to $1.21.

Houston, "hub of the oil industry" and always a boom town, was "boomier" than ever before. Thousands upon thousands of Americans rushed to the magic city from other states where the economy was declining and unemployment growing. Most of them found jobs, for plants and factories were running around the clock and new ones were being built. Towers were springing up, each one seemingly taller than the last. Demand for oilfield equipment of all kinds and sizes continued to grow and companies strove to satisfy the demand and build up inventories for an ever-brightening future.

Oil company executives, bankers, oil analysts and other experts, placed their rulers on their graphs and by extrapolation found $40 oil, $50 oil, $75 oil, $100 oil, and some, $150 oil.

As they used to say during the East Texas boom, it was like riding on a gravy train with biscuit wheels.

Along with the boom, the Petroleum Club got its first lady member—Carol Clark Tatkon, treasurer of Exxon USA. She entered quietly, as a business executive, and it was a tribute to the membership that she was accepted in the same fashion. She lunched often in the Petroleum Room with businessmen with whom she conducted Exxon business, and it was obvious from the first that she shared the pride of other members in belonging to the "finest in the land."

She was born in Seattle, Washington and named after Carole Lombard, the vivacious motion picture actress. Because her father was in government service, the Clark family moved about the

country. She went to schools in several states before being graduated from high school in Arlington, Virginia.

She was interested in the social sciences. She chose to go to Cornell University because it was a large school and coeducational.
But in her freshman year she got a taste of economics, and it
changed her life's direction. She took all of the economics courses
Cornell had to offer. She did her graduate work at New College in
New York City, then sallied forth to work on Wall Street. And she
married Daniel Tatkon, also an economic analyst.

She still had the motivations that had prompted her interest in
the social sciences, and she wanted to make money. The Tatkons
moved to Jamaica to establish an international economics consulting service where they could be of help to countries that needed it
while being paid for their expertise. The venture was a success.
But in 1963, with Jamaica in the process of dissolving its ties with
the British Commonwealth, the Tatkons decided to come home.

Her reputation preceded her. She was courted by several major
corporations, including Standard of Jersey (Exxon). She had a
question for her interviewer at Jersey: "What if I am in disagree-
ment with some company position—on proration, for example?"
"Why," said the interviewer, "we expect you to voice it or put it in
writing anytime you disagree." "Fair enough," she said.

She joined the company in 1964 as a foreign economics analyst
in the international division. (Daniel Tatkon would become a
writer and the publisher of a newsletter in the health-care field.)
By 1972, Carol was the head of the division's international finance section, and in 1973 she was named manager of the international finance planning and economics division. In 1977 she was
made manager of the corporate financial planning division of the
treasurer's department, and in 1980 she came to Houston from
New York as treasurer of Exxon USA.

During her climb up the corporate ladder Carol Clark Tatkon
had not lost her compassion, her charm, or her sense of humor.
She was a bright and welcome addition to the Club.

Howard Warren, founder, charter member and past president
of the club died during the Brock administration—on March 18,

1981. Thirty-five years had passed since that wintry day he and Wilbur Ginther had sat in their Esperson Building office and dreamed of a club for oilmen only in the oil capital of the world. It could be safely said that no one had served the Club better, or had enjoyed it more.

The special assessment Will Frank's administration had imposed on the membership brought in $315,750, and the Brock administration spent it on household improvements and furniture and equipment. Gruy's House Committee, with Broun as liaison, completed the work ahead of schedule and under budget. It was a considerable achievement. It included replacement of the pesky lateral drain lines, replacement of all light dimmers on the 44th floor, replacement of 160 armless brown chairs on the 44th, installation of stainless steel liners in four walk-in freezers and refrigerators, replacement of the rugs in the lobby and foyer, replacement of the wall covering and drapes in the Drake and Oklahoma Rooms, replacement of the charbroiler in the kitchen, and replacement of the parquet dance floor in the Petroleum Room. Replacements for the casements on the 43rd floor were ready to be installed.

The membership policy changes resulted in a decrease of corporate ownership from 77% to 70% of the total membership during the year. "Since these memberships are transferred within companies and, therefore, do not become available to the Club for new members, this is still too high, but the trend is encouraging," Brock reported to the membership. "There still remains a substantial waiting list, which reflects the high demand and the quality of our Club. . . ."

The Annual Petroleum Club Ball had been staged at hotels in the past. For the first time it was held in the Club itself. "The elegance and quality of the Ball increased greatly as a result, and at a reduced cost," Brock reported. Roxburgh, the Entertainment Committee chairman, counted almost 600 in attendance at the affair.

The Publications Committee, first under Herbert Thornton and then under Thomas Gholson, had begun work on a brochure that was completed and mailed out by the present committee chaired by Harry Jacobson. The brochure was the answer to comments by members that they wanted something to give their guests as a memento and to send to business associates to show them where planned meetings, parties, and luncheons would be held. Also printed was a floorplan of the Club to be used primarily by the staff in dealing with groups interested in holding functions at the Club.

The Finance Committee's report was brief but warming. "Increased operating revenues offset increased operating expenses," wrote Chairman John Cooney. "Dues and charges for food services remained unchanged." It was the first year in many that either dues or menu prices had not risen. But Cooney had a word of warning. "If inflation continues, it appears that some increase may be necessary in the coming year."

The "coming year," of course, would see others at the helm. At the annual meeting of May 26, 1981, Brock, Willbern, Huffington, and Armstrong retired from the Board. Elected to replace them were William Grisham, who had chaired several committees (most recently the Food Committee); Judd H. Oualline, vice-president and general manager of Getty Oil's Southern E. & P. Division; J. C. Walter, Jr., chairman of the board of Houston Oil & Minerals, Inc.; W. James Wooten, president of Texas Gas Exploration Corporation.

At the organizational meeting the next day John Jacobs was elected president. Other officers elected were Robert Stewart, first vice-president; Elbert Watson, second vice-president; H. Ross Bolton, secretary; E. C. Broun, treasurer; Thomas Stoy, assistant secretary-treasurer.

The new administration kept on board H. J. Gruy as chairman of the House Committee and John T. Cooney as chairman of the Finance Committee. William A. Carpenter was named chairman of the Food Committee, Charles A. Rosenthal was named chairman of the Publications Committee, Clinton F. Morse was named chairman of the Audit and Legal Review Committee. Carpenter was a Shell Oil Company executive, Rosenthal was director of industry affairs for Schlumberger, and Morse was with the law firm of Andrews, Kurth, Campbell, and Jones. Andre Crispin, the old reliable, was named chairman of the Wine Committee, and Kenneth W. Buckles, Senior Division Account Manager for Dowell, was named chairman of the Entertainment Committee. Director Elbert Watson was appointed to serve as chairman and liaison of the Ad Hoc Planning Committee.

Growing up in the Great Depression in the Oklahoma Dust Bowl community of Lone Wolf, John Jacobs, Jr. was certain of one thing—he would never get into the oil business. He saw too many young men with fresh degrees in geology and petroleum engineering from the University of Oklahoma taking jobs pumping gasoline—when they could get jobs at all. The oil business, he figured, was no damned good.

Jacobs' father was a Lone Wolf banker. The bank failed in 1930, and the collapse killed Jacobs, Sr. The mother moved the family to a farm.

In high school, Jacobs heard of a man in Muskogee who worked in a glass plant as a chemical enginer and made $275 a month. This man had attended a wonderful school in Georgia. Any ambitious boy with $50 and a pair of overalls could get through that school, and he could learn to be a chemical engineer.

The school was Georgia Tech Co-op School. A student went to school for three months, then worked for three months until he was graduated with a degree, if he stayed the course. Jobs were provided in those tough times by companies in Georgia, Virginia, Tennessee, and Alabama. It was the finest kind of experience because a senior might work with engineers at a Dupont Nylon plant in Virginia or engineers at a U.S. Steel plant in Birmingham.

Jacobs completed the five-year course in four years. On graduation, he was recruited by Standard of Jersey, and he went to work in the great Standard of Louisiana complex in Baton Rouge. (So much for his vow not to work for an oil company.) It was an exciting time. There was a large population of engineers working on a variety of projects—aviation gasoline, the first fluid catcracker, butyl rubber, and other petrochemicals. It was a great place for a man in his early twenties.

But Jacobs left it for Venezuela. He was a romantic then and always would be. He went to work for Creole Petroleum, the Jersey subsidiary. He became a petroleum engineer, completing wells all over the country, from Lake Maracaibo to Maturin. The Creole official staff was small and Jacobs, at 26, was rubbing shoulders with Creole greats like Jim Clark, the chief engineer who dreamed up the drilling barges to exploit Lake Maracaibo. World War II was on. Creole's goal was to produce and load into the tankers one million barrels of oil per day. The goal was reached. It was an achievement of a lifetime.

So Jacobs came home and decided to go to law school. Several accepted him but he chose Yale. He specialized in "Government Control of Business" with emphasis on the oil and gas industry. A number of New York law firms offered him jobs upon his graduation, but Jacobs headed for Dallas where two young Harvard Law School graduates offered him a berth at $300 a month. He stayed there three years.

Then, Jacobs joined forces with Harry W. Bass, a Dallas oilman, to put together a natural gas project. They acquired a lot of shut-in gas in Central Texas, most of it in big parcels, built a pipeline, put in a processing plant, and found a customer—Texas Eastern Corporation. Texas Eastern liked Jacobs' style; he was asked to become an officer of the company with headquarters in Houston. The main office was in Shreveport.

Jacobs was responsible for two notable deals. One, he bought a trillion cubic feet of gas in place—something that had never been done before—at Rayne, Louisiana. Then, he bought 3.3 trillion cubic feet of gas from Gulf under a contract that bound Gulf to provide the gas, not from just a single field as usual, but from wherever it was necessary to obtain it.

Some years earlier Texas Eastern had tried to establish a separate exploration and producing company with disappointing results. Now it was a Texas Eastern division, with lots of districts and lots of employees, but with few prospects and little hope. Texas Eastern turned it over to Jacobs.

Jacobs pared the work force from 350 to 25. They found oil in the Gulf of Mexico and elsewhere. Early on they moved into the North Sea before the play began and, with partners, found oil and gas in both the British and Norwegian sectors. They became involved with Mobil at Sable Island off Nova Scotia. Finally, with a refinery and a propane plant to handle production, the division became the largest contributor to Texas Eastern's net.

After the 1973–74 embargo, with OPEC countries in the process of nationalizing their oil industries, Jacobs' group began to worry about an oil supply if another major shortage occurred. They knew they wouldn't be able to supplement their production with purchases from the majors as in times past.

Thanks to his law school studies of government control of business, Jacobs had been able to deal handily with government entities in Britain and Norway. Now he set out to study deals and make friends with government groups in Venezuela, Nigeria, Iran, Kuwait, and Saudi Arabia. It was at a time when a company could get an option to buy government oil if it supported a government's industrialization program.

Jacobs settled on Saudi Arabia in 1975. With the Celanese Corporation and the Saudi government as partners, Texas Eastern began work on a $200 million methanol plant. Under the deal Texas Eastern gained the right to buy 50,000 barrels of oil a day. And obviously, Texas Eastern would have a head start on other companies when and if the government granted new drilling concessions.

He became a Texas Eastern director in 1976, and was vice-chairman of the board when he retired from the company in 1981. At the time of his election to the Petroleum Club presidency he had just opened an office in Houston as an international business consultant.

Jacobs was a man of ideas and many interests. He was a voracious reader, and in his ranch home near Brenham there were enough books to stock a branch library. He had a separate collection of books on the Old West and cowboys that would have delighted J. Frank Dobie.

His administration took over when the boom was at its height. The Club was full at lunch and dinner with members and guests talking business or trying to relax from it. The result was record sales of food and beverages. The Club, like the industry, was enjoying a period of unequalled business activity. But neither the industry nor the Club could shake off the albatross of inflation. Costs of food supplies and costs of maintaining a happy, healthy staff continued to mount.

It was no time to increase menu prices if the Club was to remain competitive with other clubs in the city—and below restaurant prices. John Cooney's Finance Committee and Board liaison Robert Stewart, Jr. offered some recommendations that the Board approved. The fee for new resident members was increased from $4,000 to $5,000; the non-resident fee was increased from $500 to $1,000; non-resident dues were increased from $15 to $40, and resident company transfer fees were increased from $1,000 and $1,300 to a standard $1,500.

The boom brought a concern of another kind to the administration. The towering buildings that had been erected in downtown Houston—to the north of the Club—had given the city an Andean skyline of peaks and valleys. The view from the Club's windows in that direction had been shortened and narrowed, but the scene had gained a new kind of excitement, a different grandeur.

But now there were published reports that a Canadian group planned to construct a tower of 70 or more floors directly across Bell Avenue from the Exxon Building. Such a tower would almost entirely shut off the north view, depriving the Club of a precious asset.

According to the minutes of August 11, 1981, Elbert Watson's Ad Hoc Planning Committee was asked "to investigate and report to the Board regarding proposed future construction of skyscrapers on properties adjacent to the Exxon Building, and what effect this construction would have on the Club in its present location. . . ."

Before Watson's group completed its investigation, the Canadians dropped their plans to erect the building, erasing immediate concern. But the incident recalled to mind the reality the Cawthon and Moore administrations had faced when they had negotiated with Exxon for a new lease during the great remodeling program—the cost of moving and duplicating the magnificent quarters, even if a proper site could be found, would be prohibitive.

It was still that way, it appeared.

Meanwhile, the Club had entered an agreement with Exxon to install smoke alarm and sprinkler systems in the Club. Two tragic hotel fires, one in Las Vegas, the other in Houston, had caused the Houston City Council to look hard at the city's building code.

New buildings were required to install the fire fighting equipment. But Exxon, being a good corporate citizen, had begun work on such a program before the fires and before the city's building code was strengthened. The Club and Exxon split the cost of installation.

Jacobs set up a new Pension Plan Committee chaired by J. C. Walter, Jr. to oversee the transfer of the plan's funds to the Bank of the Southwest for management and investment. The bank won out over five other financial institutions in competitive bidding, agreeing to the Board's stipulation that bank officers meet with the Board on a quarterly basis, and allow the Club to make changes in its portfolio if it desired.

William Grisham resigned from the Board and the Directors elected Ned Simes to replace him. Simes, as a former chairman of the Entertainment Committee, had wanted to see 600 in attendance at the annual Petroleum Club Ball. Now Jacobs appointed Simes liaison with Kenneth Buckles' Entertainment Committee, and at year's end both could report that more than 600 were on hand at the affair.

The membership was saddened during 1982 by the deaths of Fred Heyne, Jr. and Herbert Beardmore, Heyne on January 3, Beardmore on April 10. Heyne, Club president in 1972–73, was, it will be recalled, a charter member, and his memory was toasted by old timers in the Wildcatters Grill. Beardmore was Club president in 1960–61. He was, as Walter Sterling told an interviewer, "a real good man in every way." It was an accolade that Beardmore would have treasured.

But life went on. The House Committee, in addition to supervising the installation of the fire prevention/smoke alarm system, completed the work on the casements on the 44th floor. With Board approval, a computer billing system for the Club's accounting department was purchased. Carpeting in the Petroleum Room was replaced. Design of new chairs for the 43rd floor was commissioned. These were mundane achievements, but vital to the Club's well-being.

Charles Rosenthal's Publications Committee with Judd Oual-line as Board liaison, published an "Interim Directory" to provide an up-to-date list of Club membership which had changed by 20 percent since the Pictorial Directory had been brought out in 1980. And plans were set in motion to publish a new Pictorial Directory in 1984. Because of increased printing and mailing costs, the Committee decided to publish the "Sunburst" on a semi-annual basis. The Board accepted this decision and approved it. There was no suggestion that the magazine would be phased out, as had happened to its predecessor. As Rosenthal wrote, "Two issues should be sufficient to cover all club activities." The membership accepted it as an economic move, no more, no less.

The Wine Committee, Crispin wrote, decided to recognize some of the outstanding members who had consistently attended the wine seminars from their inception. "The insignia of the Wine Committee, consisting of a silver tastevin cup and chain, was presented to the following members for their participation and contribution to the club's Wine Seminars: Mr. Bart DeLatt and Mr. John L. Joplin." Delatt, an early member, was executive vice-president of Pano Tech Exploration Corporation, and Joplin was chairman of the board of Oil and Gas Supply Company and Associates.

The Board bowed its neck and decided to contest a ruling by the Internal Revenue Service that the Club owed $24,000 in taxes on interest income for 1977, 1978, and 1979. The IRS was wrong, the Board contended. A decision was in the future, but the Board recognized that if the IRS position prevailed, the Club would be liable for taxes on similar income for 1980, 1981, and 1982—or more if the contest dragged on. The Club's outside auditors, Pannell Kerr Foster, told the Board that as of March 31, 1982 the aggregate contingency for taxes and interest thereon for all years approximated $60,000.

At the annual membership meeting of May 25, 1982, Jacobs, Bolton, Watson, and Stewart retired from the Board. Elected to the vacancies were David M. Johnson, James R. Lesch, C. R. Palmer, and Charles D. Roxburgh. Johnson was president of

PETCO; Lesch, a former Director was by now chairman of the board, president and chief executive officer of Hughes Tool Company; Palmer was chairman of the board, president, and chief executive officer of Rowan Companies, Inc.; and Roxburgh was the Exxon executive with experience as a committee chairman.

At the organizational meeting the next day, E. C. (Ned) Broun was elected president. Also elected were Thomas Stoy, first vice-president; Thomas Feehan, second vice-president; Charles Roxburgh, secretary; David Johnson, treasurer; C. R. Palmer, assistant secretary-treasurer.

The new administration was seated at a terribly unhappy time. The gravy train with biscuit wheels had run off a cliff. The dazed survivors crawled out of the wreckage and looked about them.

The boom was over.

CHAPTER TWENTY

E. C. (NED) BROUN, as a Hughes Tool Company executive, was host to a group of drilling contractors at the Super Bowl game on January 24, 1982 (49ers 26, Bengals 21). The contractors for the most part were old-timers. As the old saying goes, they had seen fire, flood, and famine in the oil patch. And they had waxed as fat as force-fed geese during the boom. Indeed, the last week of December 1981 had seen the rig count—land rigs in operation—climb to an incredible 4,530. The highest rig count before the boom had been in the mid-1950s—2,620—when the Suez Crisis and other factors created a sudden demand for domestic production. Activity had drifted downward after that, bottoming in 1971 with only 976 rigs at work.

But now oil was bringing $38 per barrel and any piece of equipment capable of boring a hole was doing so. Still, the old-timers with Broun were a trifle concerned; the rig count had dropped a bit in the first weeks of 1982. "I'm afraid its going to drop way down," said one of them, Jim Justiss of Jena, Louisiana, a non-resident member of the Petroleum Club.

"No way," said Broun, with the rosy optimism of an executive whose plants were running around the clock.

"Want to bet?" Justiss asked.

"It'll never go below four thousand," Broun said.

"Never?"

"I'll bet you a hundred dollars it never goes below four thousand, and I'll press every fifty rigs below that," Broun said confidently.

"I'll call that bet," Justiss said.

Five weeks later, during the first week of March, Broun sent Justiss a check for $100. The rig count had dropped to 3,959. He sent other checks soon afterward; the April rig count was 3,597. But Justiss took Broun off the hook. He had donated the money to the International Association of Drilling Contractors, he wrote Broun. And he added: "You should have known better."

By the time Broun assumed the presidency of the Petroleum Club in May the rig count was 3,312, and by the time he presided over his first Board meeting in June the count stood below 3,000 at 2,931.

Week by week, month by month, the rig count fell. In April 1983—with Broun's term in office almost over—the rig count bottomed out at 1,807.

The industry, from OPEC to roughneck, was in disarray. Unemployment reached heart-breaking proportions as plants and factories closed or pared their work forces to the quick. Some companies disappeared. Some went bankrupt; there was so much talk about Chapter Eleven that a country boy might have thought it referred to the sauciest section of a red hot sex novel.

People who had fled to Houston to escape the recession in other states slowly beat their way back home while the city reeled from the shock of learning that it was not impervious to the slings and arrows of outrageous fortune. Even strong banks with wise leaders staggered under loads of oil loans, and bank examiners were busy from border to border. Service company executives like Broun looked out their windows to yards and warehouses jammed with inventories for a shrunken market.

Oil had dropped to $29 a barrel, and the world was awash in the greatest glut since the 1930s. OPEC had cut production, limited exports and lowered prices to avert a global pricing war. Gas went undelivered to buyers who had contracted with producers to "take or pay" and would not or could not do either because their markets had dwindled. Much of the offshore drilling fleet stood idle at anchor, and shipyard owners stared back and forth at unfinished rigs and cancelled orders.

If 1980 and 1981 had been the best of times, then 1982 and 1983 were the worst . . . or so they seemed.

There had been an indicator or two in late 1981 that the boom might wane, but they were weak and went unheeded. Some investors new to the oil business were complaining that the industry was drilling too many dry holes and "dog" wells that would never pay out—wells that shouldn't have been drilled in the first place. And certain Wall Streeters, who were skeptical of the industry in all seasons, began talking about $16 oil in the future. They were written off as doom-sayers by the industry and most of their brothers on the Street.

But between $8 billion and $10 billion of investment money left the industry in 1982. Most of it was new money that had made entry in 1979, 1980, and 1981. Much of it was from institutional accounts with a portion of their funds earmarked as "venture capital." So, even as these people had ventured into the oil business, they ventured out. It was somewhat like giving a man a transfusion he didn't particularly need, then slicing his carotid artery.

There were as many explanations for the collapse of the boom as there were divisions in the industry. It was obvious, however, that America had suppressed its hunger for oil. It had required 18.8 million barrels a day to satisfy the U.S. appetite in 1978. The shock of the price hikes put the country on a diet, and it managed to get by on 14.8 million barrels of oil a day in 1982.

The price hikes also contributed to the global recession that depressed oil demand in other industrialized nations and threatened to wreck the economies of some Third World countries.

Then falling oil prices prompted a switch from natural gas back to fuel oil in America and helped create a gas glut. Most in the industry called it a "gas bubble," and said that proper legislation would burst it and revive gas exploration and production. Some thought that the problem might be too complex to yield to legislative action only.

In any event, Ned Broun could aver with a grin that his assumption of the Petroleum Club presidency and the demise of the boom were strictly coincidental.

Broun was no stranger to financial disaster. He was born in Beaumont, Texas in 1923 and grew up in the biggest house in town. His father was rich from real estate, lumber, insurance and other enterprises, and Broun was driven to school by a uniformed chauffeur. But when he was 12 he stood in his front yard and watched movers carry the family furniture out of the splendid house. His father was bankrupt.

The mother was dead. Broun was sent to Houston to live in the homes of two uncles, George and Phillip Broun. Both worked for Humble.. They had joined the company when it began in 1917. Broun lived first with one, then the other. He wanted no more in life at that point than to be a petroleum engineer and work for Humble.

Broun and eight others in his graduating class at Lamar High School enrolled at Texas A&M together. It was 1941, and they knew they would be going into the army. They were in the Aggie Corps, and on December 8, 1941, they were automatically in the army. And in 1943 a group of 2,800 juniors and seniors were pulled out of school and placed on active duty.

Broun went to Officers Candidate School at Fort Sill, Oklahoma. A howitzer explosion burst both of his ear drums. He spent the rest of the war in a training cadre at Fort McClellan, Alabama. He was discharged in 1945, went back to school in 1946, and was graduated as a petroleum engineer in 1947.

Broun was interviewed by several companies, including Humble, but Humble didn't follow up. He signed on with Shell one morning, and that night Humble's personnel manager, Rand Dyer, called him at home to tell him to come to work. Broun explained the fix he was in. Dyer told him that Shell would be understanding, and Shell was. The company released its employee-for-a day, and Broun signed on with Humble.

He stayed with Humble for six years. He was district engineer at Thompson field when he left to become chief engineer for Highland Oil Company at almost double his Humble salary, an air-conditioned car and a piece of the action. Phillip Broun wouldn't speak to him for six months.

Among the salesmen who called on Broun at Highland Oil was a younger man named Clive Runnels. He was a trainee, green as grass, but he dressed as if he owned the company where he worked. He said he wanted to learn the oil business so he could go out on his own. Broun sort of took him under his wing. Runnels turned out to be the grandson of the late Shanghai Pierce, a famed rancher whose range ran from hither to yonder. There were four oilfields on the Pierce ranch.

After several years, Runnels decided he was ready. He told Broun he was going buy a half interest in a drilling company. He said he would sign Broun's note so Broun could obtain half of his interest, a quarter of the company. They would borrow some money to upgrade the rigs. Broun's contract contained an escape clause. If things were not working out to his liking after two years, he could get out at the price of the stock he went in with—about $25,000.

Things didn't work out to Broun's liking. He left Runnels in 1959. He hoped to go to work for an oil company or set up shop as a consultant. But Russell Jolley and Guy Cheesman, two old friends, talked him into going to work with them at Southwest Industries, a gas engineering and processing packaging company. He became sales manager, with stock in the company. He was executive vice-president in 1965 when Ingersoll-Rand bought the company and made it an I-R division. He became president in 1967.

He would stay at Southwest Industries more than a dozen years, but there was never a day that he didn't miss the drilling business. So it was that in 1972 he went to Dresser Industries as vice-president and manager of product development of the Petroleum and Minerals Group. Another old friend and Petroleum Club member, John Blocker, was president of the P & M Group and a corporate vice-president. Broun was Blocker's right-hand man. Blocker left Dresser in 1975, and formed Blocker Energy Corporation. Dresser then was split in two, and Broun was made president of the service groups.

(While still at Dresser, Blocker called Broun in one day. With him was a man he introduced as Tony Sanchez. Blocker had a lease map and a faded old log. He wanted Broun to examine the log. Broun demurred. "It's been fifteen years since I studied one of those." Said Blocker, "That's why I want you to look at it. It's that old." Broun studied the old log. "I'd say you've got four dry gas sands, and they probably need some stimulation." "That's what I think," Blocker said. He said he was going to take a quarter of the lease for $18,000. "I want you to take half of my quarter for nine thousand," he told Broun. Broun pleaded poverty; he had a new home and a new child. So he stayed out of the deal. The discovery well in the great Wilcox trend in Webb and Zapata Counties was drilled on the lease.)

Shortly after Broun joined Dresser he went to Russia to establish trade relations and start a sales campaign. As time passed, the Soviets told him he could improve his chances for sales if he would agree to a technology exchange. He did. He made some sales, but the technology exchange was a one-way street in favor of the Soviets. Broun developed a guarded opinion about dealing with the Russians.

He was in China six months after Nixon's visit to the country. Now American citizens travel to China and about the country in style, stay in comfortable quarters, and can get bacon and eggs for breakfast if they feel like it. But Broun's first trip was rough and unpleasant. He was searched and questioned at length in Canton. He was flown to Peking in an under-powered aircraft and fed a glass of tea and a green apple enroute. He was housed in a dormitory in Peking where armed guards patrolled the encircling wall whose top was encrusted in broken glass. Nevertheless, Broun felt more comfortable doing business with the Chinese than with the Russians. Finally, after two years of hard work, he began negotiating with the Chinese on a big deal three weeks before Christmas of 1976.

It was a tough three weeks. The Chinese delayed meetings. They cancelled some. It was obvious to Broun that they were counting on his strong desire to get back home for Christmas to get better terms from him. But Broun got stubborn. He acted as if he were ready to wait forever. The contract for the $26 million deal was signed on Christmas Day.

In December 1978 the Chinese vice-premier and other officials visited Washington. There were 37 Chinese engineers training in Dresser facilities in Houston, and the Dresser brass wanted the vice-premier and party to visit Houston as Dresser's guests. The Chinese party visited Houston, but as guests of Hughes Tool Company. ABC Television was the only media representative to have access to the party. So NBC arranged with Dresser headquarters in Dallas to interview Broun on Chinese business dealings. Broun told NBC not to film the Chinese trainees or the equipment they were training on.

The NBC interviewer pinned a microphone on Broun, and they talked in front of the camera for 15 minutes. The interviewer took the microphone off Broun and Broun began giving his secretary some instruction about a series of meetings he had on tap. The interview was over.

"Before we go," said the interviewer to Broun, "I'd like to get your opinion on trading with the Russians. You've done it."

With no microphone pinned on him, Broun gave his opinion on the subject, and it was not flattering to the Russians. The interviewer and cameraman departed. On the six o'clock news that evening, Broun saw pictures of the Chinese trainees he had said could not be filmed—and he heard himself commenting on the Russians. He knew he had twisted off at the bottom of the hole.

The Russian Embassy in Washington raised hob with Dresser headquarters and Dresser headquarters raised hob with Broun. He was taken off his job and given something else to do.

He was hurt and angry that he had allowed himself to be sandbagged. The following week he was on a quail hunting party. James Lesch, Hughes Tool's chief executive officer, told Broun, "I was really proud of what you said on TV. My wife and I were watching." Lesch didn't know about the aftermath. Broun told him. And on May 1, 1979 Broun became vice-president, market development and planning, for Hughes Tool.

In January 1980 he was promoted to executive vice-president and Group President of Production Tools and Services. Four months after he was elected president of the Petroleum Club Hughes Tool named him executive vice-president, operations, of all divisions for United States and Canada.

To help run the administration, the Directors selected the following committee chairmen: Miles Reynolds, Jr., House Committee; E. LeRoy Capps, Finance; James F. Hayes, Entertainment; Chester B. Benge, Jr., Food; A. L. Ballard, Wine. Reynolds was vice-president of Coastal Oil and Gas Corporation, Capps was vice-president of Tenneco, Inc., Hayes was a geologist, Benge was president of Big "6" Drilling Company, and Ballard was president of Ballard Exploration Company, Inc.

Clinton Morse continued as chairman of the Audit and Legal Review Committee, Charles Rosenthal continued as chairman of the Publications Committee, and Director J. C. Walter, Jr. took over the Ad Hoc Planning Committee.

Before the administration could warm its chairs, Director Wooten told the Board that he had been approached by Logan Bagby, Jr. "concerning the possibility of the Club sponsoring a skeet shoot competition at the Houston Gun Club."

Everyone knew that the former Club president and Round Table regular was a skilled fisherman and a "Dead-eye Dick" hunter. Bagby felt that there were others in the Club who felt about shooting as he did. (His wife Jean was reputed to be a better shot than Bagby. This led to a kidding remark around the Club that if "Jean would let Logan outshoot her every once in a while, he wouldn't be so hard to get along with." Bagby, of course, was easy to get along with—unless he thought his precious Club was not being run to perfection.)

Broun, who liked to pop a cap now and then himself, told Wooten to see if he could recruit a chairman to organize such a function. Wooten found one—Ronald B. Thompson, a geologist for Pel-Tex Oil Company, Inc. The skeet shoot was so successful that it became an annual event. So successful, in fact, that a later administration would be moved to bar guests at the shoot and restrict it to members only.

With the bloom off the boom, the Club's night-time party business fell off. Surprisingly, the lunch and dinner business did not. A deficit was looming, nonetheless, and Chester Benge's Food Committee, after some study, recommended that both luncheon and dinner menu prices be increased by 10 to 11 percent. Benge said the increases would enable the Club to maintain its competitiveness, would not be adverse to the membership, and would result in a projected increase sufficient to negate the estimated deficit. The Board approved the recommendation.

Broun had asked Miles Reynolds' House Committee to study the possibility of raising the limit on resident memberships as one means of keeping the budget balanced. The committee responded with a simple plan that could be implemented immediately. While the quota for non-resident members stood at 450, there were only 240 non-resident members, and no waiting list, the committee said. Why not decrease the quota from 450 to 350 and let the resident membership pick up the 100 new openings thus created, increasing the resident membership limit from 1,600 to 1,700?

Why not? said the Board. The plan was approved. And Thomas Stoy, liaison to the Membership Committee, recommended that 50 of the openings be placed in the "Producer" category with the remaining 50 placed "on hold." This also was approved.

The House Committee also was busy selecting new chairs for the Club. There is no better illustration of the dedication of the various committee members over the years than the brief report Reynolds made on chair procurement. It said: "The final selection, though difficult, resulted in a balance of utility, decor, and tradition. These chairs currently are in use in the Grill Room. Additionally, more formal Shelby-Williams wood and brocade chairs were purchased for use in the Discovery Room and Coastal Suite and further compliment the decor of these lovely rooms. . . ." These were busy men in a troubled industry, yet they gave Club business the attention that they gave their own. And the purchase of chairs was only one of many Club matters they looked after during the year.

Mrs. Eleanor Westmoreland, the Club's Director of Special Activities, retired during the year after 20 years of service. A party in her honor was attended by the Board, many past presidents and a number of Club employees. She was presented a solid gold sunburst medallion and an outpouring of affection and regard.

Broun, Feehan, and Stoy stepped down from the Board at the annual membership meeting of May 24, 1983. Elected to their seats were Frederic C. Ackman, Joe S. Farmer, and William D. Kent. Ackman was chairman of the board, president and chief executive officer of Superior Oil Company, Farmer was president and chief operating officer of Union Texas Petroleum, and Kent was president of Reading and Bates Drilling Company.

At the organizational meeting the following day, Joe Walter was elected president. Other new officers were Oualline, first vice-president; Wooten, second vice-president; Kent, secretary; Simes, treasurer. The bylaws suggested but did not demand appointment of an assistant secretary-treasurer, and Walter proposed that none be selected. The Board agreed.

Charles Rosenthal agreed to continue as chairman of the Publications Committee for the third year. C. M. Hudspeth was named chairman of the Audit and Legal Review Committee, J. J. Crowder was named chairman of the House Committee, Robert Wagner, Jr. was named chairman of the Finance Committee, J. L. Batt was named chairman of the Food Committee, Donald Keller was named chairman of the Entertainment Committee, A. Henry Lichty was named chairman of the Wine Committee. Hudspeth was an attorney, Crowder was a former Director and Entertainment Committee chairman, Wagner was executive vice-president of First City National Bank, Batt was senior vice-president of Highland Resources, Inc., Keller was president of Quintana Energy Corporation, Lichty was corporate purchasing agent for Shell Oil Company.

The new regime took over the Club's direction when its rooms were busy at lunch with members and guests who were licking their wounds and looking for a ray of sunlight through the miasma of the boom's explosion. There were few crystal balls in evidence. As Jerry Crowder, the House Committee chairman, put it, "A guy who depends on a crystal ball in this business will wind up eating broken glass."

But the oil patch breeds few sissies. For the most part, oilmen heal quickly. Roughnecks and board chairmen drink from the same pool of optimism, and both are wildcatters in their heart of hearts.

As Dad Joiner told a news reporter in the East Texas field: "Wildcattin'? All it takes is guts and acreage. It seems to help some, too, if you're smart and lucky. But the thing to remember is, a strike can make you forget your dry holes in a hurry. . . ."

The very day that Club President Joe Walter cut loose for himself in the oil business in the late 1950s, the allowable was cut to 12 days. Six months later it was cut to eight days, and oil was selling for $2.60 a barrel. Walter was inclined to remember those dates and figures in the boom's aftermath. So were many other oil and gas searchers, and the memory would help them through the Great Transition from aberrance to the road to normalcy.

So the drillships and barges began to stir at anchor. The rig count began to climb slowly and painfully out of the pit. Shell was drilling a well in the deepest water ever plumbed off the Atlantic Coast. A Gulf lease sale attracted more buyers than expected, and Getty Oil moved out quickly in its wake and made a significant discovery. So did Tenneco. A group including ambitious Diamond-Shamrock was drilling on a man-made island in Alaska's Beaufort Sea. A consortium brought in a 7,000 barrel-a-day discovery off Australia. Conoco and partners found a large pool of light oil onshore United Kingdom.

Derricks began to peek over treetops along the Gulf Coast and rise in lonely splendor on vacant prairies farther inland. In Tyler County, Texas, famed author James Michener, gathering material for a novel about Texas, was on hand when the final Schlumberger was studied on a rank wildcat drilled 14,210 feet into the Woodbine by Cliffs Drilling Company. The bit had found a gas-condensate reservoir. The field was registered with the Railroad Commission as the James Michener field. Michener considered it a signal honor.

And to the south 81-year-old Wilbur Ginther and partners were turning to the right on the Number 3 Webb County School Board, and Ginther and others were spudding in two wildcats in Webb and Colorado Counties. "Hell," Ginther told a reporter who commented on his longevity, "I'm just getting started good."

Thus, the explorers and producers moved slowly but with warming courage into uncertainty.

The service side of the industry had been the most severely wounded. It accepted as fact that it would be the slowest to recover. The survivors were lean and hungry. They hacked at their inventories like woodsmen with dull axes in a forest of ebony. But they were wildcatters, too, these service company men. They were inventors and artisans who at every step along the way from the days of Drake had provided the industry with new tools for new challenges.

CHAPTER TWENTY-ONE

JOE WALTER, JR. was a semi-rarity, a Houstonian born and reared. He was born in 1927, the son of a landman. The senior Walter and his partner, Irwin Smith, often took an override in lieu of cash commission, and from these overrides formed Houston Royalty Company. So Walter was reared in an oil atmosphere, and he went to the University of Texas where he obtained his petroleum engineering degree and a Master's in geology.

He worked for Humble for six years. He left to embark on a career that would gain him recognition as one of the most outstanding oilmen in the industry. He would call on his technical knowledge, his business acumen, his horse sense, and a natural friendliness to build a powerful independent company that was active offshore and onshore, at home and abroad. When running it ceased to be fun, he stepped out of it and went back to his beginnings. The joy to him, he realized, was in the finding of the damned stuff and building from scratch.

When Walter left Humble in the late 1950s his father had just died. He fixed up an office in Houston Royalty Company quarters and started putting drilling deals together. He was moderately successful.

Then Irwin Smith died. The stockholders wanted to sell Houston Royalty, but they asked too much to attract a buyer. Walter got a group together and bought stock control. The company had some good assets—royalties from several rich fields. Walter wanted to change it from a dormant royalty company to an exploration company.

The opportunity came when another royalty company, Royalties Management Corporation in Tulsa, put its properties up for cash bids. Without cash, Walter offered shares of Houston Royalty stock. A cash bidder won out, but the deal fell through. The Tulsa company decided to take Walter's stock merger offer. The companies together generated $700,000 a year in revenue, and provided Walter with the money to operate.

At that time the major companies were selling off their marginal properties on sealed bids. Walter studied various properties, then leaped into the bidding. He over-bid on properties he wanted; his studies had showed him that the leases might not be as marginal as they appeared to be to their owners.

The first lease he bought had 15 wells making 40 barrels per day. Three months later they were making 800. He used the new reserves to borrow money to make another deal, and he continued in this fashion, building up a formidable reserve base over the years.

In 1969 Walter induced Fox Benton Jr. to leave a large investment group and come to his side to ride shotgun for him into the future. Benton was a brilliant young financial expert. He had started out as a geologist but decided he didn't like the work. His head was full of numbers, so he went to Harvard Business School. He emerged with the numbers neatly in place and adding up to a new career.

In a flurry, the team made 18 acquisitions for $36 million. They had borrowed to the hilt, but they had increased production tremendously. So Walter went public, a giant step, mounting the American Stock Exchange as Houston Oil and Minerals (HO&M). Sales of the public offering paid off most of the debt. Life was good.

So Walter sashayed out and bought half interest in the Good Hope refinery in Good Hope, Louisiana, maker of jet fuel for sale to the U.S. Department of Defense. Four months later the bottom dropped out of the government market. Walter was stuck with his first losing proposition. He expanded the refinery. Losses increased. HO&M board of directors was growing restive.

On Christmas Day Walter's phone rang at home. Turkey and trimmings were waiting, but Walter talked on. The caller was a Jack Stanley, who owned 200 service stations. His gasoline supplier was cutting him off, Stanley said. Would Walter be interested in selling Good Hope refinery? They dickered for two hours. Walter sold to Stanley, at a profit, and enjoyed a merry Christmas indeed.

This was the first of several lucky breaks that would come his way, and Walter accepted it as such with smiling candor.

He had been high bidder on 55 Sprayberry wells belonging to Union of Texas in West Texas, but had been denied the purchase. Both sides used bad language in leaving each other. But a year later Union of Texas asked Walter if he wanted to bid on its half interest in four Miocene wells on 2,700 acres in 12 feet of water off Bolivar Point. Occidental Petroleum owned the other half interest. The wells were shut-in; they had never been produced. Both companies were highly conscious of the recent oil spill off Santa Barbara, California, and neither wanted to build production facilities so close to heavy barge traffic at Bolivar Point.

Walter bought both halves for $2.7 million. Union threw the 55 Sprayberry wells into the pot and Occidental sweetened it with six wells in Southwest Texas.

Walter had no idea of building production platforms. He leased a spoil bank called Goat Island from the state, built tanks on it, ran flowlines underwater from the wells, and began producing 150 barrels of oil per day.

There was plenty of room for other oil wells on the lease, but Walter had other ideas. Ten years earlier Texaco had plugged and abandoned a gas blowout off the structure because gas at the time was worth practically nothing. Texaco had cut five feet of Frio at 12,000 feet. Walter wanted to test the Frio on the structure. There had been no recent gas finds of significance on the Gulf Coast and the price was going up.

He drilled low on the downthrust side of a fault, and the dry hole cost $700,000. HO&M directors mopped their brows. Walter tried to sell half of the Frio to major companies to get drilling money for a second well. There were no takers. Walter finally convinced the board that HO&M had to try it on its own.

The well came in at two o'clock on a Monday morning. H. F. (Kep) Keplinger of The Keplinger Companies was called in to study the data. The bit had cut 125 feet of gas-laden Frio. Based on the one well, Keplinger came up with an estimate of from 50 billion to 100 billion cubic feet of gas. HO&M stock on Amex leaped like a frightened deer.

The gas was put out for bids. You can't look at the logs, Walter told prospective bidders. Make your bids on the basis that there's 100 bcf of gas available. If you're high bidder, then you can look at all the data. If you don't think there's 100 bcf at hand, you can get out of the deal.

Lone Star Gas was high bidder, agreeing to pay 52 cents per mcf for the gas with annual price determinations based on the highest gas contract in Railroad Commission Districts 3 and 4 . . . and with a wild card permitting HO&M to call for a determination at any time. And, since HO&M didn't have the money to develop the production, Walter got Lone Star to advance HO&M $5 million, to be repaid out of the gas flow.

(The lease eventually would support 14 oil wells and produce 500 bcf of gas.)

After that big lick, HO&M pursued the Frio all over Galveston Bay and in state coastal waters, finding both oil and gas fields in abundance, growing faster than any other independent in the business.

So HO&M went international with Will Frank (Petroleum Club President 1979–80) in command. The company held positions around the world—the North Sea, Africa, Middle East, South America. In Colombia the company bought a lease that was producing 5,000 barrels per day. The crude sold for $5.34 per barrel inside the country. Frank negotiated a deal with the government whereby the company got higher prices every time it did something to increase crude volume over normal decline. Production went up to 10,000 barrels a day, and then up to 18,000. With two new discoveries, production would reach 35,000 barrels a day. The original investment had been $55 million.

While the company was drilling on an anticline in Australia—on the way to a dry hole—the bit cut through one coal seam after another. Walter told the geologists, "Find the outcrop." They did. With a partner, Walter prepared to bid for a coal lease. The part-

ner decided to bid for itself alone, and got the lease. Walter selected another site. He went to see the state premier. He told the
premier: "I can't bid high enough alone to win a concession. But
these people who win them simply sit on them. Sell me a lease
without taking bids and we'll wake up some competition, which is
what you want. We'll agree to spend $80 million over the next five
years in development." Said the premier: "You've got it."

Development work, which included building a town for 900
people, began. In the meantime, Walter had suffered a heart attack. And he also had reached a conclusion about the coal venture. "We're explorers, not earth movers," he told Fox Benton.
Benton agreed. They sold out to a European combine for $90 mil-
lion, making about a 30 percent rate of return on the investment.
A short time later the Australian coal market sagged and fell.
Lady Luck had kissed her favorite boyfriend one more time.

Houston Oil and Minerals had not neglected the minerals business. A subsidiary, Houston Minerals Corporation, with a staff
including 85 mineral geologists, at one point held the fourth largest mineral position in the United States. Included in the holdings
were mineral properties bought from the Howard Hughes estate,
making the company owner of the townsite of Tonopah, Nevada,
distinguished for its hotel, cafe, and legal whorehouse. HMC was
involved in gold, silver, lead, and zinc operations throughout the
Old West, and with General Crude Corporation was busy with a
promising lead-zinc prospect in Alaska.

Luck was holding Walter's hand when he led the company into
the Baltimore Canyon play. The company paid $4 million for two
parcels on a flank of Stone Dome, an extrusion of igneous rock 21
miles in diameter. No one had drilled in the Canyon yet; Walter
wanted to be the first. However, another company grabbed the
lead, and the project, for Walter, lost a bit of its luster.

He was duck hunting with a Phillips Petroleum executive. The
Phillips man had just returned from New York. Bankers and security analysts, he said, seemed a bit surprised that Phillips didn't
have a Baltimore Canyon position. Said Walter, "I think I know
how you can get one."

Phillips gave HO&M the $4 million HO&M had paid for the
leases and agreed to drill two wells for a half interest in the deal.
The test was a duster.

The little company with five employees had grown into one with five separate divisions and 1,400 employees. Each division operated autonomously. Walter and Benton found themselves almost totally engulfed in personnel, pension, and regulatory problems. They had created a bunch of entrepreneurs who were having all the fun. They had had their fling at running a big company, and now it was time to get out and start over. They sought a merger partner.

Walter went to a hospital for heart surgery on April 1, 1981. He was out and frisky enough to conclude a merger with Tenneco on April 10. A week later he was turkey hunting with his doctor.

He had a ranch near Brownwood that he seldom visited. Now he went there, determined to stay until he got tired of ranch life. He stayed less than three weeks. Back in Houston he organized Walter Oil and Gas Corporation and was in full swing by autumn. The company was drilling 15 to 20 wildcats a year when he became president of the Petroleum Club in May 1983.

Walter Sterling, long-time member of the Club and its 14th president, died June 1, 1983. Lee Hill, the 26th president, died four months later on September 30. Memorial lunches marked their passage.

On a pillar near the Round Table in the southwest corner of the Wildcatters Grill is a wooden plaque bearing metallic name plates. It held the names of William J. Knight, 1974; W. J. Steger, 1974; R. C. Stuart, 1976; Carey B. O'Connor, 1980; John E. Lyons, 1980, Fred J. Heyne, Jr., 1981; Samuel B. Symington, 1981; James F. Tucker, 1983. Now the names of Sterling and Hill were added to the honored list.

The table accommodates 10 chairs comfortably, but room can always be found to squeeze in a couple of more. Perhaps 25 Club members are considered "regulars" at the Round Table, but some are more regular than others. No one is invited to "join" the Round Table, but one can tell after a few visits whether one is accepted into the brotherhood. The guest of a regular, however, is welcomed unreservedly.

It is a diverse brotherhood, diverse in age, attitude, personality. Some have been and are coarse in speech and humor to the point of ribaldry. Others have been and are smilingly proper to the point of primness. Some have been and are story-tellers, others listeners. Most of the conventional faiths are represented around the table. Still they do more than tolerate each other; they revel in each other's company.

They are men like Charles Blanchard, R. L. McVey, Grant Fuller, James Teague, Aubrey Stautberg, Don Ford, James Allison, Jr., Logan Bagby, Jr., Milton Gregory, E. Porter Johnson, C. F. (Doc) Maginnis, Eugene Shiels, Michael Kelly, C. H. (Admiral) Taylor, Ben Roshton, Chester Benge, Jr., Bruce Anderson, Delmer Bowman, Edward Bowman, E. O. Buck, Joe (Jumping Joe) King, Robert Bybee, J. J. (Jerry) Crowder, Arthur Draeger, Dan Flowers, Don McMahon, Russell Neil, Walter Plumhoff.

And S. Patrick (Father Pat) Murphy, Rector and Headmaster, St. Barnabas' Episcopal Church & School.

He appeared to be a lonely, diffident man when he walked into the Club on an August day in 1961. He was a new member, in the clergy classification. But he was not an eminent divine, as were the other nine members in that category. He was an Episcopalian priest with a small, poor parish near Hobby Airport. A friend in Garland, Texas, from whence he had been transferred, had asked a brother in Houston to get the transferee into some Club. The brother, John Baird of Hudson Engineering, had acted.

And now Father Pat looked around the Men's Grill in the old Club. There was a giant of a man sitting alone at a big table. "Mind if I sit down?" Father Pat asked.

The big man's bushy brows rose as he studied the clerical garb. "I don't mind, Sky Pilot, if you can stand the heat," said Samuel Symington. It was the least profane remark he would utter throughout the ensuing luncheon when all the chairs were occupied.

Father Pat stood the heat. He was standing it in 1981 when he preached Sam Symington's funeral. He was standing it in 1983.

Symington had not addressed a tender, naive man back there in 1961. Perhaps the collar and pink cheeks misled him. But Father Pat, then 44, had been raised an orphan. He had served his time in country towns and Chicago ghettos and college campuses. He was the kind of man bishops sent into areas to get things moving. And he got them moving.

As much as any other member of the Petroleum Club of Houston, Father Pat was a wildcatter.

The winter of 1983 was a season of optimistic doubt at the Petroleum Club. Would the Iraq-Iran war broaden and cut off crude supplies from the area? Would OPEC cut crude prices or cut production even farther and raise prices in defiance of economic law? What in the *hell* was the Reagan administration and the U.S. Congress going to do about natural gas, if anything?

The gas question was old when the Club was new. There always had been questions, and the answers had seldom if ever been satisfactory to all segments of the industry represented in the Club.

But the Club had endured. It would endure.

ACKNOWLEDGMENT

Scores of persons—Club members, employees, and others—were interviewed in the preparation of *The Finest In The Land*. They shared their memories and their records with me, and I am grateful. They made rambling through the oil patch a pleasant adventure.

I am particularly indebted to Cynthia Vann, the Club's Membership Secretary, an exhuberant lady of wit and imagination who helped me sift through files and faded clippings with the zest of a cub reporter. She rejoiced with me when some hard-to-find tidbit was unearthed and excoriated me as a dullard when one managed to escape.

I am grateful, too, to an old friend, Warren Baker, Director of the Energy Reference Service, University of Houston—Downtown. Baker "broke in" as a reporter on the old *Oil Weekly* in 1925, and spent more than 50 years in the oil magazine field as a reporter, editor, and publisher. He is as full of producing sands as the various published reports and hearings of the Multinationals Subcommittee of the Senate Foreign Relations Committee.

INDEX

317